# ANTHROPOLOGY AND AGING

# Anthropology and Aging

*Comprehensive Reviews*

*Edited by*

R.L. Rubinstein
*Philadelphia Geriatric Center, USA*

*with*

J. Keith
*Swarthmore College, USA*

D. Shenk
*St. Cloud State University, USA*

*and*

D. Wieland
*Sepulveda VA Medical Center, USA*

KLUWER ACADEMIC PUBLISHERS
DORDRECHT / BOSTON / LONDON

Library of Congress Cataloging-in-Publication Data

```
Anthropology and aging : comprehensive reviews / edited by Robert L.
  Rubinstein with Jennie Keith, Dena Shenk, and Darryl Wieland.
     p.   cm.
   ISBN 0-7923-0743-7 (U.S. : alk. paper)
   1. Aging--Anthropological aspects.  2. Aging--Cross-cultural
 studies.   I. Rubinstein, Robert L.
 GN485.A55  1990
 305.26--dc20                                              90-4376
                                                              CIP
```

ISBN 0-7923-0743-7

Published by Kluwer Academic Publishers,
P.O. Box 17, 3300 AA Dordrecht, The Netherlands

Kluwer Academic Publishers incorporates
the publishing programmes of
D. Reidel, Martinus Nijhoff, Dr W. Junk and MTP Press.

Sold and distributed in the U.S.A. and Canada
by Kluwer Academic Publishers,
101 Philip Drive, Norwell, MA 02061, U.S.A.

In all other countries, sold and distributed
by Kluwer Academic Publishers Group,
P.O. Box 322, 3300 AH Dordrecht, The Netherlands.

*Printed on acid-free paper*

All Rights Reserved
© 1990 by Kluwer Academic Publishers
No part of the material protected by this copyright notice may be reproduced or
utilized in any form or by any means, electronic or mechanical
including photocopying, recording or by any information storage and
retrieval system, without written permission from the copyright owner.

Printed in The Netherlands

# TABLE OF CONTENTS

| | |
|---|---|
| PREFACE | vii |
| ACKNOWLEDGEMENTS | ix |
| INTRODUCTION | 1 |

## SECTION ONE: BIOLOGICAL AND HEALTH ISSUES

1. DOUGLAS E. CREWS / Anthropological Issues in Biological Gerontology — 11
2. J. NEIL HENDERSON / Anthropology, Health and Aging — 39
3. BETHEL POWERS / Nursing and Aging — 69

## SECTION TWO: CULTURAL ISSUES

4. ROBERT L. RUBINSTEIN / Nature, Culture, Gender, Age: A Critical Review — 109
5. CHRISTINE FRY / The Life Course in Context: Implications of Comparative Research — 129

## SECTION THREE: AREAL STUDIES

6. CHRISTIE KIEFER / Aging and the Elderly in Japan — 153
7. ANDREA SANKAR / Gerontological Research in China: The Role for Anthropological Inquiry — 173

| | |
|---|---|
| List of Contributors | 201 |
| Combined Index | 203 |

# PREFACE

This book was conceived as a project of the Association for Anthropology and Gerontology, a multidisciplinary and international organization, formed in 1978, that is dedicated to the exploration and understanding of aging within and across the diversity of human cultures. The perspective of the Association is holistic, comparative and international. Membership is drawn from both academic and applied sectors and includes the social and biological sciences, medicine, urban planning, policy studies, social work and the development, administration and provision of services for the aged. Information about membership may be obtained from Dr. Eunice Boyer, Department of Sociology and Anthropology, Carthage College, Kenosha, Wisconsin WI, 53141 USA.

ACKNOWLEDGEMENTS

In a collective enterprise such as this, there are many people who have helped us along the way. Many members of the Association for Anthropology and Gerontology and many other colleagues gave us advice, read and commented on drafts of papers and otherwise supported this project. The editors and individual authors would like to acknowledge the following for their support and help: Baine B. Alexander, Steve Albert, C. C. Ballew, Diana Bethel, Jacob Climo, Ann Dill, Jean De Rousseau, Nancy Foner, Doris Francis, Mel Goldstein, Ralph Garruto, Tony Glascock, Charlotte Ikels, Sharon Kaufman, Jeanie Kayser-Jones, S. Loth, Mark Luborsky, Linda Mitteness, Corinne Nydegger, J. D. Pearson, David Plath, J. P. Ritchie, Phil Stafford, Rachael Stark, Maria Vesperi, Marjorie Schweitzer, Jay Sokolovsky, Toni Tripp-Reimer, Martin Whyte, and Connie Wolfsen.

ROBERT L. RUBINSTEIN

INTRODUCTION

The purpose of this volume, and what we hope will be a continuing series beginning with this volume, is to present a current accounting of critical issues and areas of interest in the anthropology of age. In particular, the events of later life as they are biologically, socially, and culturally lived have become a central concern in many societies. An awareness of national and international "graying," and its problems and potentials is widespread. Gerontology, the scholarly and practice-oriented discipline concerned with the latter segment of the life course, is unique in that it is truly multidisciplinary. Rather than the province of any one discipline, its focus is a segment of the life-span and the range of issues attendant on this period. Most gerontologists are likely to be physicians, social workers, sociologists or psychologists; however, architects, dentists, anthropologists, lawyers or nurses are also likely to be gerontologists, with interests in those areas of their topical specialty that intersect later life issues.

The presence of anthropologists in an informed and multidisciplinary study of later life is not new. However, in gerontology, anthropological research methods represent, to be sure, a minority position about the nature of the human world, about aging, and how one understands these. Yet, anthropologists are playing an increasing and increasingly vital role in the study of later life, not only because the event of graying is planet wide, international and cross-cultural, but also because of the increasing recognition of the significance of culture and cultural differences in human life.

This book then may be viewed as representing developments in four areas of the anthropological study of age. These are, an increasing interest in the anthropology of age; wider recognition of the culture concept and of anthropological research methods; the relationship of anthropological gerontology to the development of anthropology; and a strong commitment within anthropology to the examination of human life from both biological and cultural perspectives.

First, there is now a substantial group of anthropologists working on issues relating to the life-course and old age in cultural context. For example, the Association for Anthropology and Gerontology has approximately 300 members, mostly North Americans, and there are certainly more than this number with similar concerns internationally. The existence of a focal interest for this group has enabled there to be a sufficiency of current and critical work that requires timely summarization and periodic discussion in a systematic fashion. Because of its unique, hybrid position between the disciplines of anthropology and gerontology, the anthropology of aging has developed with an eye to events and issues in both disciplines.

Significantly, the level of analytic insight has increased with time. Initially, anthropologists who studied aging have occasioned three sorts of contributions to gerontology. First, they have provided interesting "case material" illustrative of the lives of elderly in other cultures. Secondly, they have acted within the overall system of social science checks and balances by providing negative examples in opposition to energetic, but inaccurate, overgeneralizations ("but my people don't do it that way at all" as Fry and Keith have noted). And third, they have provided contrasting cases for examination of hypotheses cross-culturally for example as represented in the literature on aging and modernization. Within the domain of gerontology these early efforts have generally been regarded as peripheral or distal.

But largely due to the continuing efforts of anthropologists who study later life, anthropological issues, concerns, and perspectives are increasingly central to the discipline of gerontology. As with the rest of the anthropology, anthropologists who study old age no longer work exclusively in small-scale, non-western cultures and many anthropologists who study old age are involved in research on key issues both in the West and elsewhere.

Second and certainly related to this, is the increasing recognition of the centrality of culture in human life. By necessity research from the biological perspective must treat individuals and groups as "organisms" and sort out various effects on organisms. However, if it true that much of 20th century social science can be viewed as conflict between those *social scientists* who see humans as "organisms" (who experience or are affected by an environment that is unmediated by culture or a system of cultural meaning) and those who see humans as 'suspended in a web of meaning they themselves have made', it is increasingly clear that the latter view has become the dominant one. And ironically, because of the racist overtones in biological, genetic or "organismic" social determinism, the widespread acceptance of the cultural point of view will ultimately provide the only acceptable matrix on which to rest any genetic or biological explanations for human behavior that have or will emerge.

The central role of culture in human affairs, as it has been explored in anthropology, has not always been appreciated in gerontology. For example, in the field of environment and aging, there has been a good deal of research that has portrayed the human organism as directly and in an unmediated fashion influenced by environmental events when it is clear that these events are efficacious insofar as they are perceptible and meaningful within a larger system of meaning.

Nevertheless, it appears that gerontology as a whole is increasingly coming to grips with the significance of culture in human life. And to a certain extent, along with this, there has been a greater acceptance of anthropological methods in the wider sphere of social and behavioral sciences. While it is widely acknowledged that anthropological research methods in old age research are creative, necessary to gain certain forms of knowledge (an often begrudged acknowledgement) and are colorful and inventive, many questions have been raised about their validity and the rigor of methods by mainstream social gerontologists. Because social gerontology is a hybrid discipline, and because anthropologists have sought to do work within the realm of social gerontology, they have had to describe for gerontological audiences their methods and methodological assumptions in detail, in order to attempt to communicate these to other researchers whose primary interests are "quantitative".

On a certain level, the minority status for anthropological methods is very much an issue of power politics in academic circles, of control of funding purse strings, and of key issues in the culture of social science. Cultural anthropologists and other qualitative or cross-cultural researchers have, at times, been put in the rather unenviable position of justifying to quantitative social scientists why anthropological methods and approaches are "scientific". Thus a great deal of work by been undertaken by anthropological gerontologists with the purpose of arguing a somewhat revisionist version of mainstream social gerontological gospel: towards explaining the existence of such phenomena as validity, replicability, methodological rigor, accuracy in coding and cross-checking as standard in their tool-kits.

Partially because of the hybrid nature of the discipline of gerontology, partially because of other gerontologists ultimate desire to include anthropologists, and partially because of the politics of research funding in the United States, anthropologists of aging and other "qualitative" researchers who study later life have been responsible several excellent and worthwhile publications on anthropological research methods in the study of later life and so there has been recent productive growth in accounts of anthropological methodology in old age research. In American gerontology, the work of anthropologists has been generally subsumed in a discourse that concerns "qualitative research." This is a category that for better or worse includes the variety of forms of anthropological fieldwork and ethnographic interviewing, work by ethnographic and grounded theory sociologists, clinical psychologists who rely on case studies, and the assorted foray by humanists. However, there are few, if any, areas of anthropology that can boast such a successful array of publications on method.

All anthropological research methods, whether from a social, medical, or biological perspective ultimately derive from and make reference to the experience of fieldwork and participant observation in a culturally alien setting. Further, a variety of methods (derived from the psychological centrality of this experience in anthropology) are now utilized in old age research.

When anthropologists carry out cross-cultural participant observation or fieldwork, (the epitome of qualitative research) they rely on such intimate strategies as coresidence in the community of study and long periods of continual association. In a home society, or when working on a specific problem or topic (such as "caregiving"), an anthropologists may rely more heavily on key informant interviewing (rather than free flowing participation and residence) in their work.

The relationship of these two sorts of projects, which certainly make up the bulk of research projects conducted by anthropologists who study old age, may be conceptually unclear.

In her well-known paper on participant observation in old age research, Jennie Keith has outlined three stages in carrying out participant observation in old age research. An examination of these is helpful, I believe, in understanding the typological relationship of research strategies. These research stages represent movement from the greatest outsideness to the greatest insideness. Stage One, the initial period, is topically all inclusive; the researcher asks as many questions of as many of her subjects as possible. Stage Two consists of a more focused, discriminating inquiry. In this stage the researcher has gained enough general background knowledge of a setting to enable her to investigate in detail specific areas of interest. Data collection becomes more specialized. In Stage Three, there occurs a further refinement of data collection and the posing and testing of hypotheses (for those who view research as a hypothesis testing endeavor). Additionally, in Stage Three, there is (hopefully) growing language competence in a cross-cultural setting and increased personal closeness with key informants that permits the researcher to gain the greatest degree of insight into informants' subjectivities and the inner workings and meanings within the culture.

This outline helps in gaining a perspective on the relationship of traditional anthropological participant observation in other cultures to same-culture projects that consist primarily of ethnographic key informant interviewing. That relationship is this. Cross cultural field research consists of the three stage structure. In same-culture research, the first two stages may drop out, and the ethnographer, in essence, begins at Stage Three, or late Stage Two at the earliest, having enough language competence, background knowledge, and a good deal of working knowledge to begin at a more focused level.

Third, this work builds on long-standing strengths and world-views of anthropologists. From the original cross cultural project on the status of older people by Simmons in the 1930's, the anthropology of aging has examined such topics as modernization in cross-cultural or cross-national perspective; aging in American culture and the meaning of independence; death hastening; the experience of health and illness and the effects of medicalization and the culture of medicine on the elderly; the meaning of disability; the influence of specific environments such as nursing homes, old age residences, single room occupancy hotels or particular cities or communities on elders; the meaning of retirement; and aging and ethnicity. More recent works have focused on aging in a diversity of settings; the analysis of language, narrative, and meaning; gender roles; the meaning of giving and helping; lives of minority aged; caregiving; and on specific and comparative cultural studies. Concern with methodology has also been important. Two recently established journals, the

Journal of Cross-Cultural Gerontology and The Journal of Aging Studies have viewed anthropological input as central.

Further, research on aging by anthropologists speaks to a continued interest in areal studies that has been central in academic anthropology from the beginning of the 20th century. In one sense, the key focus in the properties of culture areas, their similarities, features and historical circumstances that have made them distinctive, has a very real manifestation in the domain of applied anthropology. The original interest in culture areas was diagnostic and organizational and under Boasian conception related to a theory of culture process. In places such as native California, for example, an interest in culture area was also reactive: sorting things out before it was too late, before all the traditional native cultures and the worlds they represented were gone. However, the present interest in the properties of culture areas, in one sense, relates to the amelioration of human suffering and the enhancement of human dignity. Recognizing the cultural differences that are a part of human diversity, (and given the increasing interest in the centrality of human culture that was mentioned above) we then ask questions like these: "What is old age like in such-and-such a setting? How does it relate to cultural and other structural features? Is life for the aged better there than it is here? What can we learn from them? What are their medical systems and notions of health and well-being and how do people there utilize them?"

Culture has generally been defined as the distinctive properties of a historically-constituted group of people, the system of community values and collective roles and behaviors. But if culture is the meaning content and the meaning content is mediated by symbols (as is widely recognized) then culture need not be defined in this way. Culture may also be conceptualized as a system of shared symbols that give meaning to experience. Thereby the distinctive property of an institution, and ethnic group, or any other collectivity or setting that provides a backdrop for distinctive meaning also constitutes a culture. Nursing homes, senior centers, ethnic groups and nations all have and share cultures. And an emerging focus in the anthropology of age examines personal culture: the realm of personal meanings routines and ideas and the relationship and process of relationship of these to wider, community values.

Notions of separate cultures, the distinctiveness of cultures, the uniqueness of cultures, the inviolability of the culture as an analytic unit have had a deservedly important role in western anthropology.

As was noted above, the single culture paradigm is still the major focal unit in cultural analysis although other cultural levels are equally significant and, additionally, there have been numerous suggestions that other cross-cutting devices such as class and gender are more effective analytical edges for certain purposes.

It is true, too, that an examination of aging might have much to gain from the perspective of the world system, for example by applying insights from the type of analysis developed by Eric Wolf in *Europe and the People Without*

*History*. This work sought to make clear the global process of economic and social change and interaction beginning around 1400 AD that profoundly influenced the structure of all world societies. This set of occurrences, Wolf believes, flies in the face of the anthropological image of distinctive, rather pristine, cultures, the essentialist equation of one society with one culture, and a still existing bias, in many quarters, against the study of pluralism.

A fourth aspect of this book is that it sustains and renews the commitment of anthropology to an approach that examines human life from a perspective of both biology and culture. Both implicitly and explicitly, all of the chapters below, although situated from one of the other perspective, raise questions that are germane to the other. Their mutuality is clear.

Two important notions have bound anthropology together in all of its diversity. The first is the culture concept and the notion of the centrality of culture in human life. While this is gaining renew vigor and new acceptance throughout the social sciences, ironically it has come under attack from within anthropology. Related to this is the traditional view of anthropology as constructed of four quadrants, physical and cultural anthropology, archaeology, and linguistics. This book represents a commitment to this paradigm and in particular to an anthropology of age that is constituted of both biological and cultural perspectives.

The fact that this relationship must be stated and reiterated has much to do with our perception of boundaries. This is a continual problem in human sciences and derives not from any natural or given set of boundaries or entities but rather from the culturally based predisposing factors that lead us to categorize the world in the ways we do.

Analytically, marking boundaries has been a central problem in the holocultural or comparative study of societies and of the elderly within them: when in both time and space, does one society end and another begin?

And this too has been a central problem in ethnographic individualism or person-centered ethnography: How do we honor the individual as a unit both separate from but comparable to other such units? How do we separate the culturally or self-constructed aspect of the individual as a unit of study from the biological or any externally motivated influence? Is the separation of biology and culture, or the reduction of culture to biology or of biology to culture a result of the western construction of the person and of causality?

As we explore in increasing depth the biological, psychosocial and cultural dimensions of the aging individual, it is increasing clear that personal constructs and themes are complex, rich, and unique. Do our methods adequately capture these so that they are not reduced to other than what they are?

The papers that follow in this volume represent key aspects of anthropology: its concern with human diversity, with the essence of humanity, and with human well-being. The papers are diverse, representing critical analyses and reviews that range from biological anthropology, nursing and aging, and health and aging, through more culturally situated issues, the life course in comparative perspective and gender and age, to a precise focus on two cultures and nations in looks at aging in China and Japan. It is our hope that future volumes will look at anthropological contributions to other topics and areas.

In concluding this introductory essay, I would like to note what I feel to be three goals in the anthropological study of old age that are variously represented in the content and spirit of the essays.

The first goal is knowledge for knowledge sake. This may seem old fashioned but to put it another way: it is only with contextually accurate knowledge of ourselves that we can address our human condition. We all know now that anthropology has had a colonial heritage, in large part derived from the colonial context of status and power asymmetry or has relied heavily on status or gender inequality between researcher and informant, and in some cases has served colonial interests. We all know of continuing efforts to "rebirth" anthropology in a modern context. In this regard, we must continue our efforts to understand the relationship between knowledge and power. We must use analytical tools that enable us to ask the question, Is there any way we can perceive old age and the social and biological contexts of old age, for anything other than what we "normally" take these to be?

A second goal is the elucidation of meaning. Simply put, the anthropological perspective on later life is the only tool for understanding meaning. There are no others. The relationship of the material conditions of life to these is only somewhat causal, it seems to me, and these conditions are mediated by systems of perception. The examination of these phenomena is what the anthropologist does best.

A third goal in the anthropological study of old age is social advocacy and the education of awareness. Without this out work becomes socially meaningless. While some social sciences and scientists attempt a value-free stance, anthropologists by and large recognize the presence of values in their work and it is these human values that inform both their selection of research methods and color their results.

# SECTION ONE

# BIOLOGICAL AND HEALTH ISSUES

DOUGLAS E. CREWS

# 1. ANTHROPOLOGICAL ISSUES IN BIOLOGICAL GERONTOLOGY

INTRODUCTION

Biological anthropologists have contributed substantially to the development of growth standards and measures of biological age. They have also examined morphological and physiological variation and change during adulthood. Nevertheless, relatively little research has focused on the later stages of human life.

This paper highlights areas of gerontological research that historically have been investigated by biological anthropologists. The topical approach reflects evolutionary and ecological methodologies which have been described as "human population biology" (Baker 1982) and my own interests and biases. Evolutionary perspectives, factors influencing human longevity, characteristics of human aging, and areas for future research are examined in turn. The focus is biological anthropology, although results from other disciplines are freely included in keeping with the transdisciplinary nature of human population biology. Materials reviewed recently by other biological anthropologists (Borkan, Hults and Mayer 1982; Beall 1984), theories of aging, as well as rates of decline in specific functions and organs, are not discussed in detail. This information is available elsewhere (Beall 1987; Bergsma and Harrison 1978; Bittles and Collins 1986; Comfort 1979; Finch and Schneider 1985; Mayer 1987; Moore 1987; Warner *et al.* 1987).

"Aging" includes all time dependent structural and functional changes, both maturational and senescent, that normally occur in the postpubertal period among males and females of a species (Bowden and Williams 1984). Senescence is a characteristic unique to living systems. It is a progressive, irreversible, cumulative functional deterioration, that likely occurs secondary to evolutionary inertia of complex biological systems. Disease is a deviation from normal aging and is associated with decreased adaptability and performance (Plato 1987). Longevity is the duration of individual life, while life expectancy is a statistical probability statement, the mean length of life remaining to members of a defined population. Life span is the maximum length of life experienced by a member of a defined population in a particular environment (see Mayer 1987; Moore 1987).

## EVOLUTIONARY PERSPECTIVES

*Evolution of Human Longevity*

Biological anthropologists generally approach the evolution of human life span via three routes: evolutionary theory, comparisons with nonhuman primates, and skeletal evidence. Life expectancy and life span increased during consecutive stages of hominid evolution (Cutler 1976; Timeras 1972; Williams 1957). Some have suggested that old age may be a recent evolutionary development attained only by modern humans (Dolhinow 1980). However, old nonhuman primates exhibit many features during aging that are homologous to those observed in humans (Hrdy 1981).

Based on probable ancestral body size, maximum life spans of 30-40 years have been estimated for protohominids of about 4-5 million years ago (MYA) (Dolhinow 1980). Weiss (1981) estimated 42 years for Ramapithecus (14 MYA) and 51 years for Australopithecus (3 MYA). Life span apparently increased to 71 years in Homo erectus and to 93+ years in Neanderthals and modern humans (Cutler 1976). Weiss (1981, 1984, 1989) estimated average life expectancies of 15 years for Australopithecines, 18 years for Neanderthals, 25 years for Neolithic Agriculturalists and Classic/Medieval Europeans, 43 years during the 19th century, 55 years in the early 20th century and over 75 years in contemporary cosmopolitan settings. These estimates would suggest that estimated maximum life spans of 40 years for later Neanderthals (Trinkaus and Thompson 1987) are low.

*Longevity and Menopause*

It has been suggested that separation of senescence from cessation of reproduction may be an uniquely human trait (Lancaster and King 1985). The fact that captive chimpanzees appeared to menstruate into their fifth decade of life lent support to this hypothesis (Bowden and Jones 1979). However, others report that both in captivity and in the wild, macaques and chimpanzees show a post-reproductive stage during late life similar enough to provide a model for human menopause (Gould, Flint and Graham 1981; Graham, Kling and Steiner 1979; Hogden *et al.* 1977; Hrdy 1981). These findings suggest that menopause may be an integral aspect of higher primate reproductive strategies in response to aging, a possibility in need of further research.

Menopause is associated with a number of age-related changes including increased risks of breast cancer and cardiovascular diseases, bone loss, fractures, falls, and associated disability, and changes in circulating hormones (Lancaster and King 1985; Mayer 1982; Moore 1981). Nevertheless, menopause and the extended post-reproductive survival of humans remains an enigma. Lancaster and King (1985) examined ethnographic data from modern hunter-gathers. Their findings indicate that 53% of women who survive to age

15 will live to age 45, suggesting that post-reproductive survival likely occurred among Neolithic and Paleolithic populations. They proposed an evolutionary hypothesis for menopause based on lactation amenorrhea being the normal biological state during most of the life span of most women who have ever lived. Among prehistoric women increased birth intervals and lactation amenorrhea gradually extended to what today is recognized as menopause. For most of human evolution this state likely occurred within a hormonal context of lactation, while a women was still nursing her last-born child(ren), possibly as extended lactation amenorrhea.

Menopause is variable in its age of onset. Worldwide the median age ranges from 43 to 51 years (Beall and Weitz 1989). This range provides ample variation to investigate genetic, ecological, and environmental relationships. In a high-altitude Tibetan population menopause occurs at the early mean age of 47 years, compared to a Western standard of 51 years. Age at menarche on the other hand is delayed, occurring at a mean age of 16 years (Beall 1983). Among !Kung Bushmen of the Kalahari desert menarche occurs at mean age 17 and menopause at 40 years (Howell 1979). Women in such populations have shorter reproductive spans compared to women living in cosmopolitan societies where menarche occurs as early as age 12 and menopause after 50 years (Lancaster and King 1985). Ages at menarche and menopause are associated with other physiological features. Women who experience earlier menarche have more body fat (Lancaster and King 1985), and earlier maturers may be more prone to develop breast cancer (Micozzi 1987). Furthermore, at least among Nepali women, shorter and leaner women experience earlier menopause (Beall 1986). If such differences are associated with differential fertility or mortality this may lead to selection for larger body size and greater body fat (see Beall 1986).

*Skeletal Evidence*

Evolutionary theories of human longevity are dependent on fossil materials. Unfortunately, skeletal assemblages usually do not reflect the true age-sex structures of the populations which produced them (Molleson 1986; Walker, Johnson and Lambert 1988). Numerous sources of bias and inaccuracy accompany the aging of skeletal materials. Nonrandom factors, including time, place of death, and cultural practices, have resulted in differential disposal of the dead (Molleson 1981, 1986; Workshop of European Anthropologists 1980; Sattenspiel and Hardpending 1983), while differential preservation leads to loss of elderly adult remains (Walker, Johnson and Lambert 1988).

The rarity of Neanderthal fossil materials estimated to have been over 45 years at death led some investigators to the hypothesis that postreproductive survival was absent in this group (Trinkaus and Thompson 1987). Others have suggested that the lack of estimated ages over 60 years in historic skeletal samples result because aging techniques are insensitive (Molleson 1986). Recently, systematic biases have been revealed in skeletal aging methods that use the pubic symphysis (Jackes 1985). Similarly, the Todd and the McKern-Stewart aging methods do not allow for age-related variability (Katz and Suchey 1986).

Age estimates of skeletal remains have always been difficult, particularly in adults (Iscan 1988). New methods to estimate ages at death have been developed using the male os pubis (Suchey and Brooks 1987), cementum annulation in human teeth (Charles et al. 1986; Condon et al. 1986), microscopic examination of cortical bone and teeth (Kerley and Ubelaker 1978; Maples 1978), and morphological metamorphoses at the sternal rib end, a method which can be accurately used into the seventh decade (Iscan, Loth and Wright 1984a, 1984b). Such methods may provide more accurate age estimates to examine the evolution of human longevity than traditional archaeological and forensic methods (Jackes 1985). New methods using the pubic symphysis or the auricular surface of the ilium may also be compromised because validation studies were based on mixed sex/race samples with poorly documented ages (Iscan 1988). Given the variety of cultural filters, temporal variation in skeletal features, and historic and scientific biases leading to preservation and recovery, reservations will always attend the estimation of demographic parameters from skeletal samples (Howell 1982; Loth and Iscan 1988).

Present estimates of life span among fossil hominid populations are based on two methods: skeletal aging and body/brain size regression. Estimates from body and brain size regression may be more reliable since they are not hindered by problems underlying skeletal estimates (Weiss 1981). Furthermore, these methods have been validated across mammalian species and predict hominid life spans as well as those of other mammals. However, they are based upon skeletal remains, that are themselves biased by preservation, recovery, and estimating techniques for physique and brain volume. Still these methods indicate relatively stable life span from the Neanderthal through the modern stage of human evolution and suggest that no special factors need be invoked to account for longevity of modern humans (Weiss 1981, 1984).

INFLUENCES ON HUMAN LONGEVITY

*Growth and Development*

Early growth and development are likely to influence age-related disease onset and ultimate longevity. Anthropologists have examined both biological and environmental influences on growth and development. For instance, Clark and colleagues (1985, 1986, 1988, 1989) examined poor early neurological and thymus development and found them to be associated with adult immunodysfunction and subsequent longevity. They also confirmed that serum levels of thymosine-alpha 1, a master regulator of a variety of immune functions, were associated with poor early growth and shorter life span. Such research suggests neurological and immunological factors occurring as early as fetal development and infancy influence differential senescence. Apparently, proxy measurements for some such impairments may be observed in skeletal materials. These and additional markers may provide clues to differential longevity and survival in historical and paleontological samples.

Longevity may be affected adversely by early maturation. Early maturity is associated with a persisting centripetal fat pattern that is associated with chronic diseases of adulthood (Frisancho and Flegel 1982). Across national populations early maturity is associated with increased breast cancer rates (Micozzi 1987). Interestingly, late maturity apparently is associated with more complex dermatoglyphic patterns (Meier, Goodson and Roche 1986). If dermatoglyphic patterns, traits that show high heritability, do produce indicators of timing of maturity, they also may allow early identification of persons at greater risk of early development of age-related chronic diseases and consequent premature mortality as they have for several childhood diseases (see Hoff, Garruto and Durham 1989). Validation of biomarkers indicative of developmental differences is needed to better understand the influence of individual variation in disease onset and longevity.

*Aggregation of Life Span*

Pedigree analysis suggested that longevity is genetically transmitted and heritable in humans (Pearl 1931; Pearl and Pearl 1934). This relationship has been reported to be strongest between a woman and her offspring (Abbot et al. 1974; Pearl and Pearl 1934). However, in the Duke First Longitudinal Study of Aging, father's age at death was the more significant predictor of longevity differences for both men and women (Palmore 1982). Recent results from a study of longevity in Mennonite families showed a higher correlation for the same sex parental relationships, that is mother-daughter or father-son, than for opposite sex relationships (Koertvelyessy et al. 1982). In general, it appears that correlations between longevity of parents and their children are weak (Swedlund et al. 1983; Vaupel 1988).

Indications are that levels of frailty or susceptibility to specific diseases rather than longevity *per se* may be inherited (Vaupel 1988). For instance, study of almost 1,000 adopted children in Denmark revealed that their risk for premature mortality (death prior to age 50) was double if one of their biological parents had died prematurely (Sørensen et al. 1988). Risks of death from cardiovascular diseases and infections were 4.5 and 5.8 times higher if a biological parent had died of either of these causes prior to age 50 than if neither had. Twin and sib studies show closer correlation of longevity among twin pairs than among non-twin sib pairs; twins also show shorter longevity than their non-twin sibs (Carmelli 1982). These studies suggest that environmental factors, constitutional frailty and/or fetal development may be more influential in longevity of twins. Examination of the genetic inheritance of general susceptibility to acute and chronic diseases may lead to improved understanding of biological differences in longevity.

Greatly extended life spans in some populations have never been confirmed. Instead, age exaggeration has been documented for most claims of exceptional longevity (Beall 1987; Mazess and Mathisen 1982). For example, Soviet mortality and census data for 1959 through 1979 were used to document inaccuracies in reports of extreme longevity and numbers of centenarians, and "phenomenal longevity" ended in Abkhazia when the last exceptionally long-lived cohort, i.e. those with the most inaccurate ages, died (Bennett and Garson 1986).

*Genetics and Life Span*

Genetic factors may be associated with either shorter or longer life. As much as 7% of the human genome may modulate various aspects of aging, but it is possible that only a few dozen genes are of major importance (Martin 1987). Regulatory genes rather than structural genes may be involved in longevity differences and heterozygosity may be associated with greater longevity than homozygosity (Johnson 1988). Crawford and coworkers examined this hypothesis in a Mennonite community with use of eight blood group loci and ability to taste phenylthiocarbamide (PTC) as measures of heterozygosity. No association between blood groups or PTC-tasting and longevity were observed (Crawford and Rogers 1982; Koertvelyessy, Crawford and Hutchinson 1982; Rogers and Crawford 1981). Although this indicates several known blood polymorphisms are not biomarkers of aging, it still remains likely that heterozygosity in regulatory genes, membrane receptors, or metabolic or repair proteins influence disease and longevity.

In experimental organisms aging is related to changes in mRNA, decreased cell proliferation, lags in enzyme action, and loss of immune function. One factor which may account for these changes is DNA-repair enzymes, which are correlated with life spans in insects and rodents (Lamb 1986). These factors are currently being investigated in human cells. Study of DNA damage and repair in older passage human fibroblasts *in vitro* suggests that proliferative life span is neither accompanied by nor caused by accumulation of DNA strand breaks or the cell's capacity to rejoin DNA breaks (Mayer, Bradley and Nichols 1986a). At the same time, blood lymphocytes from older donors repair double-strand breaks in DNA less than half as fully (28%) as do cells from younger donors (66-78%) (Mayer, Bradley and Nichols 1986b). Apparently some aspects of aging may be related to a progressive loss of DNA repair capability. However, DNA repair shows variable age-related patterns in different cell types and at different temperatures (Mayer, Bradley and Nichols 1987). DNA repair represents a complex system and its effects on longevity are not yet fully understood. At present, it appears that an age-related decline in the fidelity of DNA reproduction may be secondary to less efficient repair of damaged sections.

*Nutrition*

Diet is one of the most potent modulators of mammalian life spans (Masoro 1988; McCay, Crowell and Maynard 1935; Roth, Cutler and Ingram 1988). Biological anthropologists share a tradition of research on dietary essentials and adequacy in children and adults but have only recently begun to examine needs of the elderly (Armelagos 1987; Ryan *et al.* 1989; Stini 1987), an issue which has been treated in depth elsewhere (Moment 1982; Ordy, Hartman and Alfin-Slatr 1984; Rivlin and Young 1982). Dietary requirements for essential nutrients have not been determined for elderly individuals (Young and Scrimshaw 1979). Studies which show age-related declines in calories, fat, saturated fat, and cholesterol intake may indicate average dietary change, while poorly estimating the range of variation, nutrient availability, or absorbtion (Elahi *et al.* 1983). Some research suggests that daily energy intakes of U.S. elderly may be as low as two-thirds of recommended daily allowances (RDA) (Bowman and Rosenberg 1982). However, in New Mexico, among healthy free-living men and women aged 60 and older, energy intakes approached 90% and 87% of RDA, with only 11% and 16%, respectively, of respondents reporting less than 100% of RDAs for protein intake (Garrey *et al.* 1982). Individual variations in requirements may be important determinants of age-related disease patterns and senescent changes. The finding that nutrient intakes of older men are more influenced by living arrangements and income than are those of older women suggests that sociocultural factors may partly explain differential nutritional patterns (Ryan *et al.* 1989).

    Comparison of anthropometric data from the 1960-62 Health Examination survey (HES) and the 1971-74 Health and Nutrition Examination Survey (HANES) revealed secular trends toward larger girths in both sexes (Bishop, Bowen and Ritchey 1981). Confounding of secular and age-related changes in anthropometric measurements suggests that current sex-specific norms for nutritional assessment using anthropometric methods may not be valid (Bishop, Bowen and Ritchey 1981). Biochemical measures such as those available to assess vitamin status, are preferable to dietary information for measuring nutritional status in elderly patients (Garrey and Hunt 1987). Longitudinal studies of nutritional status, dietary adequacy, and anthropometric traits in the elderly are necessary to develop biochemical indicators of nutritional status and to determine sociocultural and dietary interactions (Chumlea, Roche and Rogers 1984a; Garrey *et al.* 1982; Ryan *et al.* 1989).

    Quantitative studies documenting the range of cross-cultural variation in diet among the elderly, particularly in developing societies, may help in the development of new standards for dietary adequacy in later life (Katona-Apte 1984). Present day dietary standards suggest that as many as 73% of males and 23% of females may be undernourished among aged Zulus of South Africa (Ndaba and O'Keefe 1986). However, this may be an artifact of the scale used to measure. In elderly Guatamalans nutritional status was inversely related to body composition and height, factors that are associated with age, reproductive spans, and chronic diseases (Siu *et al.* 1987). Cross-cultural studies allow examination of sociocultural and nutritional interactions, however long-term biomarkers of dietary intakes and adequacy are needed to replace questionnaires which do not reveal lifetime patterns or allow comparative study (Kohrs and Czajka-Narins 1986).

Caloric restriction leads to retarded growth, development, and aging in rodents (McKay, Crowell and Maynard 1935). Although these relationships continue to be explored, the mechanism for life extension remains unknown (Moment 1982; Weindruch 1984). Life extension in rodents is not due to decreased intake of a specific nutrient or toxin, nor to delayed maturation or growth retardation, nor to decreased metabolic rates (Masoro 1987; Merry 1986).

The applicability of life extension by dietary restriction in rodents to humans has yet to be determined (Armelagos 1987). Measures of nutritional status observed in calorie restricted rodents and associated with improved longevity tend to also be associated with continued normal immunological function in elderly men and women, suggesting that the immunosuppression of aging may be partly related to diet (Goodwin and Garrey 1988; Weindruch *et al.* 1988). In captive macaques high fat or low protein diets lead to reduced longevity (Short, Williams and Bowen 1987). In an obese human sample, an isocaloric reduction diet, low in fats, led to improvements in cardiovascular risk factors and lower estimates of biological age (Kinner and Ries 1986). Food restriction and life span may be correlated in humans and other primates. Walford (1983; 1986) has recommended a calorie restricted high fiber diet to extend longevity in humans. Use of caloricly restricted but nutritionally adequate diets provides a possible method for noninvasive nutritional restriction studies in humans to document ages of onset for established age-related disease in restricted and non-restricted individuals.

## CHARACTERISTICS OF HUMAN AGING

Particular interests of biological anthropologists include changes in fat distribution, skeletal mass and content, muscle mass, blood components, and immunity that accompany aging. Human biologists are contributing to aging research in these areas by examining morphological and biochemical changes indicative of normal aging processes in diverse ethnic groups and populations.

### Body Composition

Body composition shows consistent and substantial change over the human life span (Forbes 1976; Rogers 1982; Stini 1984).
Using computerized tomography to examine age-related changes in body

composition, Borkan and coworkers (Borkan and Hults 1983; Borkan, Hults and Glyn 1983a; Borkan *et al.* 1983b; Borkan *et al.* 1985) found greater fat infiltration into abdominal and leg lean tissue, less subcutaneous and more intra-abdominal fat, and reduced muscle mass in the arms and legs of older compared to younger men. Age changes in fat patterning show cross-cultural similarities and variations. In an age-stratified sample of older Canadians, with 1.6% of men and 0.9% of women over age 65, age was associated with greater central and upper body fat, masculinization (Mueller *et al.* 1986). In Papua New Guineans fat distribution in women becomes more masculine with age but no increase in fatness was reported, in fact fatness actually decreased (Norgan 1987). Masculinization, redistribution of fat from the extremities to the trunk, the majority of which appears to occur during the second through fourth decades of life, may be a universal phenomenon of human aging (Mueller 1982; Mueller *et al.* 1986). However, healthy samples representing normal aged persons are still lacking, and the continuity of observed patterns across latter years of the lifespan has not been fully established.

Vague (1956) and Feldman, Sender and Siegelaub (1969) were among the first to relate differences in fat patterning to risks of chronic diseases. An area of continued epidemiological and anthropological concern (Björntrop, 1987; Mueller *et al.* 1986; Vague *et al.* 1985). Among participants in the 1960-62 Health Examination Survey an increasing upper body distribution of fat was evident from ages 18 to 79 years (Gillum 1987). Higher values for upper body fat in turn were associated with higher blood pressure, higher post-load serum glucose, and higher serum cholesterol levels. The masculinized fat pattern has been reported to be associated with arteriosclerosis (Vague *et al.* 1985), hypertension (Gillum 1987), blood glucose and diabetes (Mueller 1982; Newell-Morris 1989; Szathmary and Holt 1983, Vague *et al.* 1985) and coronary heart disease (Gillum 1987; Lapidus *et al.* 1984; Larsson *et al.* 1984), all age-related pathologies. The need for lifespan analysis of fat patterning to detect epidemiologic and genetic associations beyond the sixth decade of life is obvious. Computerized tomography provides a method for determining age-related change in body composition and fat distribution (Borkan, Hults and Glyn 1983a; Borkan *et al.* 1985; Seidell *et al.* 1987). However, computerized tomography is expensive and the equipment is not portable, which limits applicability in determining epidemiological associations.

Lean body mass and muscle mass also show age-related declines in adults which may lead to lower metabolic rates and concomitant declines in energy requirements with age (Rogers 1982; Stini 1984). This may partly explain the lower consumption of calories, carbohydrates, and fat by elderly men and women than young men and women on a nonrestricted free-choice diet (Wurtman *et al.* 1988). Among United States men, arm circumferences increase until 35-44 years and then steadily decrease; in women arm circumferences and triceps skinfold thicknesses increase until 45-65 years and then may stabilize or decline with age (Bishop, Bowen and Ritchey 1981). Muscle strength, muscle area, and strength exerted per unit muscle area are lower in older than in younger individuals (Pearson, Bassey and Bendall 1985). Cross-culturally elderly Welsh men and women have less fat and muscle than their United States counterparts, but their fat and muscle volume still decline with age (Burr and Philips 1984). The extent of cross-cultural variation in muscle mass declines with age remains to be fully documented. Such age-related declines in muscle mass likely are related to the accelerating loss of neuromuscular performance observed with age.

In U.S. Mennonites neuromuscular performance for six traits declined an average of 43% by age 85, relative to the level achieved at age 30, the majority of this loss occurred after age 55. This pattern of aging may generalize to the entire human neuromotor system (Devor, Crawford and Osness 1985). Since skeletal muscle mediates immune response and recovery from trauma, life span may reflect inherent differences in individual capacity to recover from stress at different ages (Stini 1984). The majority of age-related decline in muscle and neuromuscular performance have been confirmed in cross-sectional studies, extension of these studies cross-culturally and longitudinally will improve understanding of adaptive processes that influence longevity.

Anthropometric measurements, skinfolds, girths, body mass index, arm circumference, arm muscle circumference, and frame size are used to assess nutritional status, health, and age-related body remodeling (Frisancho 1974; Stoudt 1981; Vir and Love 1980). Since these measurements show larger interobserver errors and greater variance in older than in younger adults (Chumlea *et al.* 1984a, 1984b, 1986), accuracy of measurement becomes of greater absolute and relative importance with age. Skinfold measurements appear to be more accurate estimates of subcutaneous fat in the elderly, than do ultrasound measurements (Chumlea and Roche 1986; Shepard *et al.* 1985). However the two methods apparently are little different for estimating body density (Jones, Davies and Norgan 1986). Bioelectrical impedance, a relatively new method for determining body composition may be useful in this arena and practical for field and clinic use (Baumgartner, Chumlea and Roche 1989).

*Skeletal Mass and Composition*

Progressive bone loss may be a cross-culturally valid universal of human aging. Patterns of adult bone loss vary between and within populations and by sex, but the majority of age-related variation remains unexplained. In most samples, loss of skeletal compact bone begins between the third and fifth decades of life. Women ultimately lose two to three times as much bone as men, with the largest sex differential occurring post-menopausally (Mazess 1982; Plato 1987; Roche 1966). Although women using estrogen exhibit higher cortical thickness and lower medullary widths than those not on estrogen, it has not been demonstrated that either calcium supplements or physical activity can reverse, although either may delay, age-related adult bone loss (Plato 1987).

Age-related cortical bone resorption at the endosteal surface leading to enlargement of the medullary space and thinning of the cortical ring is well-documented (Ericksen 1982; Garn 1975; Rundgren, Eklund and Jonson 1984). Age-related cortical bone loss was observed in both cross-sectional and longitudinal studies of hand-wrist radiographs from the Baltimore Longitudinal Study of Aging (BLSA) (Fox, Tobin and Plato 1986). Between ages 45 and 69 years a 2% per decade loss of cortical bone was associated with age, but was not associated with cohort differences or secular trend. Age-related bone loss occurs at different times, different rates, and to different degrees throughout the human skeleton (Cummings *et al.* 1985; Fox, Tobin and Plato 1986; Garn 1975; Plato 1987). This is partly due to trabecular bone loss beginning a decade earlier than cortical bone loss and the two types of bone being differentially distributed at various sites (Cummings *et al.* 1985; Mazess 1982).

Osteoporosis is a pathological age-related condition of bone loss often associated with but clinically distinct from osteoarthritis. Among Belgium women of similar age (50-75 years) and skeletal structure those with osteoporosis were shorter, more slender, and had less fat, and muscle strength, and smaller muscle girths; those with osteoarthritis had more fat, were more obese, had greater muscle mass and muscle strength than women without either condition (Dequekar, Goris and Uytterhoeven 1983).

In a study of United States women including individuals less than 50 to over 80 years of age, osteoporosis was associated with age-related increases in the incidence of hip fracture, and hip fracture was in turn associated with a 15-20% increase in one-year mortality (Cummings *et al.* 1985). The higher frequency of osteoporotic fractures in the trochanteric region (50%), distal forearm (50-70%), and vertebrae (>66%) were probably account for by the larger amounts of trabecular relative to cortical bone at these sites. Cummings and coworkers suggest that age-related hormonal declines, in 1,25-dihydroxy-vitamin D, and constitutional changes in the osteoid are likely to be more important in the osteoporotic process than estrogen, calcitrophic hormones, or diet (Cummings *et al.* 1985).

Cross-cultural studies show that although skeletal mass and bone density decline in all populations with age, individual patterns are highly variable. Among African whites aged 5 to 75 years bone density increased more rapidly with age during maturation than it did in African blacks (Solomon 1979). In African blacks bone density declined more rapidly after age 40 than in African whites, although the incidence of femoral fractures did not increase as it did in whites. Others have reported that ribs of American blacks do not show the same amount or degree of thinning in old age as do those of American whites (Iscan, Loth and Wright 1984a, 1984b). After adjusting for years since onset of menopause, bone density also declines more rapidly in North

American Indians than in U.S. white samples of postmenopausal women (Evers, Orchard and Haddad 1985). Even between closely related samples variable patterns of bone loss are observed. In two Punjab villages, Indian women showed different patterns of age-related stature decrease (Sidhu, Singal and Kansal 1983). In Tuscon and Sun City, Arizona, bone mineral was apparently retained among men 70 years and older, but rates of bone loss differed among women of the same age in the two locations (Stini 1983). No mechanisms have been postulated to explain such differences in bone loss with age in both closely related and widely divergent samples. Comparative studies which emphasize measurement techniques and longitudinal data may be necessary to determine sources of variation.

Nonhuman primates provide animal models to study the age-related development of degenerative joint disease, osteopenia/osteoporosis, and joint mobility (DeRousseau 1985a, 1985b). Degenerative joint disease, bone loss, and atrophy at joint surfaces are features of macaque as well as human aging. Age-related cortical bone loss in macaques appears to be similar to that observed in human aging (Williams and Bowden 1984), suggesting that age-related osteopenia may be a general primate characteristic (DeRousseau 1985b). However, a study of 11 recovered chimpanzee skeletons from Gombe National Park showed a lack of osteophytosis and infrequent evidence of age-related degenerative changes (Jurmain 1989). Environment and behavior interact to influence age-related skeletal changes in captive nonhuman primates (DeRousseau, Bito and Kaufman 1987; Turnquist 1983). Research with varying enclosure sizes, free-ranging and penned nonhuman primates suggests that habitual activity patterns produce differential stresses on the skeleton and influence patterns of bone loss. Back pain and vertebral body changes seen in humans, a condition almost ubiquitous by age 40, may be related to an aspect of habitual activity patterns, bipediality, that differentiates humans from hominoid apes (see Jurmain 1989).

*Biochemical Factors*

Associations of blood cholesterol and glucose with age and increased risks of chronic diseases suggest that nutritional biochemistry and molecular biology provide more than one window to examine aging processes. Age-related increases in blood glucose and the consequent increase in non-enzymatic glycation of macromolecules may be another universal of human aging (Andres and Tobin 1974; Cerami 1986; Reaven and Reaven 1985). The main glucoregulating hormones, insulin and glucagon, are not associated significantly with age, however both are positively associated with obesity, and obesity is often independently associated with plasma glucose and age (Elahi *et al.* 1982). Among participants in the Baltimore Longitudinal Study of Aging, age, obesity, and diet were each associated with glucose tolerance in the elderly (Tobin 1987). In a sample of 66 adult Samoans age was associated positively and significantly with blood glucose and glycated hemoglobin, but body mass index, skinfolds, and percent trunk fat were not (Crews and Bindon 1989). Glucose, specifically the consequent high non-enzymatic glycation of proteins and nucleic acids that occurs in a high glucose environment, has been suggested as a possible explanation for some age-related changes documented in extracellular proteins and cells (Cerami 1986). This theory was recently shown to be compatible with what is known of longevity extension in the food-restricted rodent model (Masoro, Katz and McMahon 1989) and needs to be examined in epidemiological studies of human aging.

Progressive age-related declines in immunocompetence and an increased frequency of autoimmune disorders with age have suggested that changes in immune function may be fundamentals of human aging. Both thymus and bone marrow mediated immunity decline with age in mice and humans (Makinodan and Yunis 1977; Walford 1983; Walford et al. 1974). These declines may be partly mediated by diet (Walford 1986), antioxidant levels (Tengerdy 1980), DNA repair (Mayer et al. 1987), and/or developmental factors (Clark et al. 1988, 1989). Defects in proliferative capacity of cultured peripheral T-lymphocytes have been reported in old and diabetic men and patients with Down's syndrome (Franceschi et al. 1982). These defects occur at younger ages in diabetic men and very early in Down's syndrome patients, who are immunodeficient throughout life. However, wide variability in measurable immunoglobulins with age has generally characterized studies in divergent populations and may suggest local selective factors rather than true age-related change. In Guamanian Chamorros total lymphocyte and T-cell counts are negatively correlated with age, while IgA increases, IgM decreases, and IgG shows no cross-sectional pattern of change between ages 20 and 83 years (Hoffman et al. 1983). However, in Quechua Indians IgA and IgM increase until middle age and then decrease, while IgG and IgD increase during growth and development, but remain stable through adulthood (Memeo et al. 1982). Such wide variation may indicate differences in local ecological or historical circumstances or the presence of disease vectors rather than age-related losses or changes.

In U.S. elderly, immune function declines and evidence of subclinical malnutrition increase with age, but the two trends are apparently not associated with each other (Goodwin and Garry 1988). This may indicate that immunosuppression is secondary to aging itself. However, increased life span in dietary restricted rodents is not associated with Thymosine-alpha 1 levels (Weindruch et al. 1988), which suggests that changing immune competence may be disease related rather than a normal processes of aging. Conversely, administration of Thymosine-alpha-1 appears to enhance immune function in aging nonhuman primates (Ershler et al. 1988).

*Neurobiology*

With increasing age, neurological degenerative diseases become more common. Reviewing the etiology and neuropathology of Alzheimer's Dementia, Mann (1985) suggested that the primary cause may be leakage of a neurotoxin into the cerebral compartment. Garruto, Gajdusek, and their colleagues (Gajdusek 1963, 1984; Garruto 1989; Garruto and Gajdusek 1985; Garruto et al. 1983; 1984; Garruto, Yanagihara and Gajdusek 1985; Yanagihara et al. 1984) have examined several foci of dementia-type diseases. Their work in West New Guinea, Guam and the Northern Mariana Islands suggests that aluminum, in combination with silicon and calcium, may be neurotoxins as postulated by Mann. High levels of aluminum in the local soil combined with low levels of calcium and magnesium appear to have led to chronic deposition of minerals in central nervous system tissues (Garruto et al. 1985). A consequent defect in mineral metabolism apparently leads to active transport and deposition of these elements into brain and spinal cord tissues. Subtle abnormalities in calcium metabolism, including elevated parathyroid hormone levels and reduced cortical bone mass, along with aluminum, silicon, and calcium deposits are associated with amyotrophic lateral sclerosis and Parkinson dementia in Guamanian Chamorros. On Guam such abnormalities in calcium metabolism are lifelong and occur prior to the onset of disease (Yanagihara et al. 1984; Garruto 1989).

The neurofibillary degeneration seen in Guamanian patients with amyotrophic lateral sclerosis and Parkinson dementia appear histologically and morphologically similar to those of patients with Alzheimer's disease (Garruto 1989). These same neuropathological features of Alzheimer's disease are seen in almost all Down's Syndrome patients after age 45 (Heston 1984). The same neurofibillary tangles seen in Alzheimer's disease and Down's syndrome also occur in multi-infarct dementia, a large number of patients with Parkinson dementia, and in the normally aging brain (Wright and Whalley 1984). This suggests that the neurofibillary degeneration seen in Guamanian amyotrophic lateral sclerosis, Parkinson dementia, Alzheimer's disease, and Down's syndrome may be exaggerations of normal processes of neuronal aging and may be related to a genetic defect in mineral metabolism which leads to selective deposition of cations in neuronal tissues. Pedigree analysis has demonstrated a higher incidence of Down's syndrome and Alzheimer's disease among relatives of probands dying from Alzheimer's disease (Heston 1984), suggesting a genetic component to Alzheimer's disease. The available evidence links accelerated neuronal senescence with a genetic factor, possibly located on chromosome 21 (Heston 1984; Schweber 1985). Recent results confirm that the precursor protein for the beta amyloid protein constituent of the neurofibril tangles seen in Alzheimer's disease is located on chromosome 21 (Joachim and Selkoe 1989). The precursor protein has been highly conserved in evolution and although located on chromosome 21 apparently is excluded from the minimal region known to be associated with Down's syndrome phenotype. In addition this beta amyloid protein appears to be glycated (Joachim and Selkoe 1989).

*Demographic Changes and Models*

In the year 2000 A.D. approximately 61% of the world's population aged 65 years and older will reside in developing societies (Hoover and Siegal 1986; Torrey, Kinsella and Taeuber 1987). In many cultures and social systems unaccustomed to large numbers of elderly, increases of 292% (Guatemala) to 357% (Brazil) in the proportion over 65 may occur; in others only small (Sweden: 21%) and moderate (United States: 105%) increases should occur (Torrey et al. 1987). Life expectancies, which now range from 49 (Bangladesh) to 77 years (Japan), may approach 85 years by 2000 A.D. in some contemporary cosmopolitan nations. It has been suggested that little further change will occur in expectation of individual life as survival curves become more rectangular (Fries 1980; Fries and Crapo 1981). This proposition has been questioned (Weiss 1989).

As methods for estimating debility and frailty in the elderly have developed, it has become clear that the age at which functional degeneration and infirmity begin have increased during the latter half of the 20th century in several contemporary societies (Manton 1986a, 1986b; Stout and Crawford 1988). Manton used multiple-cause mortality data to show that frailty among older Americans decreased between 1968 and 1980. Use of similar methods indicated that debility had not yet had an opportunity to increase nor decrease among aging residents of American Samoa between 1950 and 1980 (Crews 1989). A survey of English hospital records suggested that active life expectancy increased between 1954 and 1986, but at the same time age of onset of disability or of long-term hospitalization and the length of terminal dependency increased (Stout and Crawford 1988). These results indicate social, emotional, and medical costs associated with the prolongation of life that are not easily quantified.

Although patterns observed today were probably more pronounced prior to the industrial and agricultural revolutions, mortality crossovers provide data to examine the interaction of physiological and cultural influences on life span (Markides and Machalek 1984). Mortality crossovers at later ages have been observed between black and white segments of the United States population (Nam, Weatherby and Ockey 1978; Wing et al. 1985). Population heterogeneity and selective survival have been offered as explanations for these crossovers, which occur in almost all populations composed of advantaged and disadvantaged racial or social segments (Manton, Poss and Wing 1979; Manton and Stallard 1984). In disadvantaged segments, selective survival may result in unexpectedly high numbers of healthy elderly who, although they may live longer, may ultimately require less medical care.

## FUTURE RESEARCH

Available data suggest that the human life span is finite (Fries 1988), perhaps due to inherent properties of biological systems (Rose and Graves 1989). Longevity can apparently be enhanced through nutritional and perhaps other means. However, it is unlikely that genes coding specifically for senescence have evolved (Kirkwood and Holiday 1986; Warner et al. 1987). What have evolved are mechanisms, including DNA repair enzymes, cell turn-over, and immunological factors, which reduce the effects of inherent declines and external insults and thereby enhance longevity. Constants or universals of human aging are likely to include increased susceptibility to degenerative and infectious diseases, fat redistribution and infiltration into muscle tissue, progressive loss in the ability to repair damaged DNA, declining immune competence, neuronal degeneration, elevated plasma glucose with a correlated increase in glycated macromolecules, and loss of bone and muscle mass. Other changes commonly seen with increasing age at present may be classified as disease. Biological anthropologists are in an unique position to distinguish between progressive changes secondary to aging and those associated with disease (Pearson and Crews 1989).

Although present environments are not likely to reflect those in which most human populations evolved (Weiss 1989), biological anthropologists view much of human physiologic variability as the result of adaptation to various environments. If longevity represents evolutionary compromises between individual reproductive success and the evolutionary success of genes (Rose and Graves 1989), this adaptive perspective should be a valuable paradigm for studies of human senescence (Mayer 1987).

Using a comparative approach contemporary biological anthropologists and human population biologists are in the position to use their unique perspectives on demography, adaptation, secular trends, and human variability to integratively examine relationships of biomedical and sociocultural factors with human aging and to determine those features that are uniquely human aging, part of our primate heritage, and/or general mammalian patterns. As a group they are trained to document change and variation, determine sources of variation through group comparisons, and explore functional and behavioral consequences of observed variation (Beall and Weitz 1989). This comparative perspective is particularly important in the description and definition of "normal aging." Relationships between "average" and "normal" aging remain unclear, partly because variation in all quantitatively measured factors increases with age. Hypotheses based on evolutionary paradigms and tested using comparative methods help differentiate between aging and disease and aid in determining rates of aging in different environments and populations. A focus on healthy normal aging as opposed to average aging leads to comparative studies. Representative samples stratified by only age, sex, and/or ethnicity confound age-related changes with disease and are of little utility in determining normal aging. Standards developed using healthy individuals without known disease who show little or no functional losses may be more beneficial and would aid in evaluating the validity of proposed relationships and the utility of interventions hypothesized to retard aging processes.

An integrated human population biology approach has seldom been used to examine relationships among age, social, and physiological factors. Partial examples include Beall and Goldstein's (1986) work with Nepali populations (Beall, Goldstein and Feldman 1985), Garruto and coworkers' (1989) study of isolated foci of Alzheimer-type dementias in the Pacific, and Crawford and coworkers' studies among U.S. Mennonites (Crawford and Rogers 1982). Longitudinal studies in anthropological populations that use a life-course perspective are necessary to document the extent and nature of age-related variation and ages of onset of decrement in physiological functions in genetically different populations residing in various cultural and environmental settings (Pearson and Crews 1989). This will improve understanding of health and physical functioning among elderly in all cultures, and of one of the primary problems facing gerontological research in the next century, differentiating disease from normal aging.

Life span, post-reproductive survival, and menopause should be of particular interest to biological anthropologists interested in human evolution. Genetic models, including reproductive success and inclusive fitness may provide new hypotheses to examine aging and senescence (Hamilton 1964). Still, the subjects of senescence, longevity, and the long post-reproductive survival of *Homo sapiens* may frequently be overlooked by anthropologists studying the evolution of uniquely human characteristics (e.g. Hardpending, Rogers and Draper 1987).

The suggestion that longevity may be associated with heterozygosity and specific genetic traits is intriguing. Trisomy 21 is definitely associated with reduced longevity, however the influence of other commonly examined genetic traits on longevity are either relatively small or not yet determined. Laboratory animals can be used to examine genetic, morphological, and physiological variation. However, only human studies help determine how variation in environment, sociocultural factors, genetics, and stress influence longevity. Furthermore, it is possible that mechanisms of aging and senescence differ between phyla, families, and even genera and that specific studies will ultimately be necessary at each level (Rose and Graves 1989)

Today the aged are a large and increasing proportion of many populations. The consequences of this basic change in relationships between the old and the rest of society have yet to be elucidated (Eisdorfer 1981). Sociocultural definitions of "old age" have not kept pace with the reality of maintenance of continued high physiological function into the sixth, seventh, and eighth decades. Analyses of cohort trends in per cent surviving show the aged to be an ever older population segment. Definitions of old age originally were related to decreasing physical and mental function. These definitions have changed little in recent decades. However, many functional capabilities have undergone secular trends in ecological situations less confounded by childhood diseases, environmental insult, and poor nutrition.

Scientific and social thinking about long-livedness presents a problem in cultural evolution (Weiss 1981). When life expectancy averaged 30 years, living 50 years may have been exceptional. When life expectancy averages 75 years, 65 is not remarkable. To determine long-lived and the segregation of this trait, pedigree and population genetic studies of aging require precise definitions. Since aging and senescence are poorly represented by chronological age, redefinition of such concepts will benefit gerontological studies in general. Although biological age estimates tend toward a functional framework, present definitions of youth, maturity, and old age also need amendment. Average life expectancy often exceeds common definitions of old age. For example, over 80% of the United States population lives long enough to be eligible to receive social security benefits. Referring to persons over age 85, the fastest growing segment of many populations, as the "oldest old" (Suzman and Riley 1985) does not clarify differences between a 65 year old with a biological age of 86 and an 86 year old with a biological age of 65. The option to combine all persons over age 65 into a single category no longer exists, but adequate alternatives have not been fully developed. Research suggests that as life expectancy and active life expectancy have increased, debility has decreased in some industrial populations during the twentieth century. Significantly, the duration of terminal dependency, although postponed, has also increased (Stout and Crawford 1988), and may be a cause for social and economic concern.

Beall (1984) suggested that a clearer understanding of human variation might be achieved if biological anthropologists focused on age as an organizing principle. She urged that adaptive capacity and variation in later life be studied explicitly using sophisticated models already developed and validated among youth and adults. To date, a number of biological anthropologists have followed this paradigm. Current research indicates that more are focusing their attention in this direction.

## NOTE

This work was partly supported by a post-doctoral fellowship in Cardiovascular Epidemiology, Biostatistics, Nutrition, and Prevention (NHLBI #5T32HL07113) and National Institute on Aging Grant #RO1 AG08395 to D.E. Crews. This chapter was written at Department of Community Health and Preventive Medicine, Northwestern University Medical School.

## REFERENCES

Abbot, M.H., E.A. Murphy, D.R. Bolling and H. Abbey (1974) The Familial Component in Longevity: A Study of Nonagenarians II. Preliminary Analysis of the Completed Study. Johns Hopkins Medical Journal 134:1-16.

Andres, R. and J.D. Tobin (1974) Aging and the Disposition of Glucose. In Advances in Experimental Medicine and Biology, vol. 61. V. Cristofalo, T. Roberts, and R. Adelman, eds. New York: Plenum Press.

Armelagos, G.J. and L. Sibley (1987) Diet and Longevity: Myth and Reality. American Journal of Physical Anthropology 72:175.

Baumgartner R.N., W.C. Chumlea and A.F. Roche (1989) Estimation of Body Composition from Bioelectrical Impedance of Body Segments. American Journal of Clinical Nutrition 50:221-226.

Baker, P.T. (1982) Human Population Biology: A Viable Transdisciplinary Science. Human Biology 54:203-220.

Beall, C.M. (1983) Ages at Menopause and Menarche in a High Altitude Himalayan Population. Annals of Human Biology 10:365-370.

Beall, C.M. (1984) Theoretical Dimensions of a Focus on Age in Physical Anthropological. In Age and Anthropological Theory. D.I. Kertzer and J. Keith, eds. Ithaca, NY: Cornell University Press.

Beall, C.M. (1986) Factors Associated with Menopause Status in a High Altitude Tibetan Population. American Journal of Physical Anthropology 69:174.

Beall, C.M. (1987) Studies of Longevity. In The Elderly as Modern Pioneers. P. Silverman, ed. Bloomington, IN: Indiana University Press.

Beall, C.M. and C.A. Weitz (1989) The Human Population Biology of Aging. In Human Population Biology: A Transdisciplinary Science. M.A. Little and J.D. Hass, eds. Oxford: Oxford University Press, 189-200.

Beall, C.M. and M.C. Goldstein (1986) Age Differences in Sensory and Cognitive Function in an Elderly Non-western Population. Journal of Gerontology 41:387-389.

Beall, C.M., M.C. Goldstein and E.S. Feldman (1985) The Physical Fitness of Elderly Napalese Farmers Residing in Rugged Mountains and Flat Terrain. Journal of Gerontology 40:529-535.

Bennett, N.G. and L.K. Garson (1986) Extraordinary Longevity in the Soviet Union: Fact or Artifact. The Gerontologist 26:358-361.

Bergsma, D. and D.E. Harrison eds. (1978) Genetic Effects on Aging. New York: Alan R. Liss, Inc.

Bishop, C.W., P.E. Bowen and S.J. Ritchey (1981) Norms for Nutritional Assessment of American Adults by Upper Arm Anthropometry. American Journal of Clinical Nutrition 34:2530-2539.

Bittles A.H. and K.J. Collins eds. (1986) The Biology of Human Ageing. Cambridge: Cambridge University Press.

Björntorp, P. (1987) Fat Patterning and Disease: A Review. In Human Body Composition and Fat Patterning. N.G. Norgan, ed. Euro-Nut Report No. 8. Wageningen, Holland: tEuro-Nut.

Borkan, G.A. and A.H. Norris (1980) Assessment of Biological Age Using a Profile of Physical Parameters. Journal of Gerontology 35:177-184.

Borkan, G.A. and D.E. Hults (1983) Change in Body Fat Content and Distribution During Aging. American Journal of Physical Anthropology 60:175.

Borkan, G.A., D.E. Hults and P. Mayer (1982) Physical Anthropological Approaches to Aging. Yearbook of Physical Anthropology 25:181-202.

Borkan, G.A., D.E. Hults and R.J. Glynn (1983a) Role of Longitudinal Change and Secular Trend in Age Differences in Male Body Dimensions. Human Biology 53:629-641.

Borkan, G.A., D.E. Hults, S.G. Gerzof, A.H. Robbins and C.K. Silbert (1983b) Age Changes in Body Composition Revealed by Computed Tomography. Journal of Gerontology 38:673-677.

Borkan, G.A., D.E. Hults, S.G. Gerzof, A.H. Robbins and C.K. Silbert (1985) Comparison of Body Composition in Middle-aged and Elderly Males Using Computed Tomography. American Journal of Physical Anthropology 66:289-295.

Bowden, D.M. and D.D. Williams (1984) Aging. In Advances in Veterinary Science and Comparative Medicine. C.E. Cornelius, C.F. Simpson and A.G. Hendrickx, eds. New York: Academic Press.

Bowden, D.M. and M.L. Jones (1979) Aging Research in Nonhuman Primates. In Aging in Nonhuman Primates. D.M. Bowden, ed. New York: Van Nostrand Reinhold.

Bowman, B.B. and I.H. Rosenberg (1982) Assessment of the Nutritional Status of the Elderly. American Journal of Clinical Nutrition 35:1142-1151.

Burr, M.L. and K.M. Philips (1984) Anthropometric Norms in the Elderly. British Journal of Nutrition 51:165-169.

Carmelli, D. (1982) Intrapair Comparisons of Total Life Spans in Twins and Pairs of Sibs. Human Biology 54:525-537.

Cerami, A. (1986) Aging of Proteins and Nucleic Acids: What is the Role of Glucose? Trends in Biochemical Sciences 11:311-314.

Charles, D.K., K. Condon, J.M. Cheverud and J.E. Buikstra (1986) Cementum Annulation and Age Determination in Homo sapians: I. Tooth Variability and Observer Error. American Journal of Physical Anthropology 71:311-320.

Chumlea, W.C., A.F. Roche and E. Rogers (1984a) Replicability for Anthropometry in the Elderly. Human Biology 56:329-337.

Chumlea, W.C., A.F. Roche and P. Webb (1984b) Body Size, Subcutaneous Fatness and Total Body Fat in Older Adults. International Journal of Obesity 8:311-317.

Chumlea, W.C., A.F. Roche and D. Mukherjee (1986) Some Anthropometric Indices of Body Composition for Elderly Adults. Journal of Gerontology 41:36-39.

Chumlea, W.C. and A.F. Roche (1986) Ultrasonic and Skinfold Measures of Subcutaneous Adipose Tissue Thickness in Elderly Men and Women. American Journal of Physical Anthropology 71:351-357.

Clark, G.A. (1985) Heterochrony, Allometry, and Canalization in the Human Vertebral Column: Examples from Prehistoric Amerindian Populations. Ph.D. Dissertation, Department of Anthropology, University of Massachusetts, Amherst. International Microfilms, #8509532.

Clark, G.A., N.R. Hall, G.J. Armelagos, G.A. Borkan, M.M. Panjabi and F. T. Wetzel (1986) Poor Growth Prior to Early Childhood: Decreased Health and Life-span in the Adult. American Journal of Physical Anthropology 70:145-160.

Clark, G.A., N.R. Hall, C.M. Aldwin, J.M. Harris, G.A. Borkan and M. Srinivasan (1988) Measures of Poor Early Growth are Correlated with Lower Adult Levels of Thymosin-alpha-1. Human Biology 60:435-451.

Clark, G.A., C.M. Aldwin, N.R. Hall, A. Spiro, and A. Goldstein (1989) Is Poor Early Growth Related to Adult Immune Aging?: A Follow-up Study. American Journal of Human Biology 1:331-337.

Comfort, A. (1979) The Biology of Senescence, third edition. New York: Elsevier.

Condon, K. D.K. Charles, J.M. Cheverud and J.E. Buikstra (1986) Cementum Annulation and Age Determination in Homo sapians: I. American Journal of Physical Anthropology 71:321-330.

Crawford, M.H. and L. Rogers (1982) Population Genetic Models in the Study of Aging and Longevity in a Mennonite Community. Social Science and Medicine 16:149-153.

Crews, D.E. (1989) Cause Specific Mortality, Life Expectancy and Debilitation in Aging Samoans. American Journal of Human Biology 1:347-353.

Crews, D.E. and J.R. Bindon (1989) Age, Glucose, and Fat Patterning in an Obese Population. The Gerontologist 29 (Special Issue):225A.

Cummings, S.R., J.L.Kelsey, M.C. Nevitt and K.J. O'Dowd (1985) Epidemiology of Osteoporosis and Osteoporoyic Fractures. Epidemiologic Reviews 7:178-208.

Cutler, R.G. (1976) Evolution of Longevity in Primates. Journal of Human Evolution 51:169-202.

Dequekar, J., P. Goris and R. Uytterhoeven (1983) Osteoporosis and Osteoarthritis (osteoarthrosis). Anthropometric Distinctions. JAMA 249:1448-1451.

DeRousseau, C.J. (1985a) Aging in the Musculoskeletal System of Rhesus Monkeys: III. Degenerative Joint Disease. American Journal of Physical Anthropology 67:177-184.

DeRousseau, C.J. (1985b) Aging in the Musculoskeletal System of Rhesus Monkeys: III. Bone Loss. American Journal of Physical Anthropology 68:157-167.

DeRousseau, C.J., L.Z. Bito, P.L. Kaufman (1987) Aging and Life Cycles in Non-human Primates. Paper presented at the 40th Annual Scientific Meeting of the Gerontological Society of America. The Gerontologist 27 (Special Issue):12.

Devor, E.J., M.H. Crawford and W. Osness (1985) Neuromuscular Performance in a Kansas Mennonite Community: Age and Sex Effects in Performance. Human Biology 54:197-211.

Dolhinow, P. (1980) The Primates: Age, Behavior, and Evolution. In Age and Anthropological Theory. D.I. Kertzer and J. Keith, eds. Ithaca, NY: Cornell University Press.

Eisdorfer, C. (1981) Forward. In Other Ways of Growing Old. P.T. Amoss and S. Harrel, eds. Stanford, CA: Stanford University Press.

Elahi, D., D.C. Muller, S.P. Tzankoff, R. Andres and J.D. Tobin (1982) Effect of Age and Obesity on Fasting Levels of Glucose, Insulin, Glucagon, and Growth Hormone in Man. Journal of Gerontology 37:485-491.

Elahi, V.K., D. Elahi, R. Andres, J.D. Tobin, M.G. Butler and A.H. Norris (1983) A Longitudinal Study of Nutritional Intake in Men. Journal of Gerontology 38:162-180.

Ericksen, O.O. (1982) Aging Changes in the Thickness of the Proximal Femoral Cortex. American Journal of Physical Anthropology 59:121-130.

Ershler, W.B., C.L. Coe, N. Laughlin, R.G. Klopp, S. Gravenstein, E.B. Roecker and K.T. Schultz. (1988) Aging and Immunity in Non-human Primates: II. Lymphocyte Response in Thymosin Treated Middle-aged Monkeys. Journal of Gerontology: Biological Sciences 43:B142-B146.

Evers, S.E., J.W. Orchard and R.G. Haddad (1985) Bone Density in Postmenopausal North American Indians and Caucasian Females. Human Biology 57:719-726.

Feldman, R., A.J. Sender and A.B. Siegelaub (1969) Difference in Diabetic and Nondiabetic Fat Distribution Patterns by Skinfold Measurements. Diabetes 18:478-486.

Finch, C.E. and E.L. Schneider (1985) Handbook of the Biology of Aging. New York: Van Nostrand Reinhold.

Forbes, G.B. (1976) The Adult Decline in Lean Body Mass. Human Biology 48:161-173.

Fox, K.M., J.D. Tobin and C.C. Plato (1986) Longitudinal Study of Bone Loss in the Second Metacarpal. Calcified Tissue International 39:218-225.

Franceschi, C., F. Licastro, M. Chiricolo, M. Zannotti, N. Fabris, E. Macchegiani, L. Tabacchi, F. Barboni, M. Masi (1982) Selective Deficiency of T-lymphocytes Subset(s) in Aged and in Down's Syndrome Subjects. In Developments in Hematology and Immunology, Volume 3: Immunology and Aging. N. Fabris, ed.

Fries, J.F. (1980) Aging, Natural Death and the Compression of Mortality. New England Journal of Medicine 300:130-135.

Fries, J.F. and L.M.Crapo (1981) Vitality and Aging: Implications of the Rectangular Curve. San Francisco: W.H.Freeman and Co.

Fries, J.F. (1988) Aging, Illness and Health Policy: Implications of the Compression of Mortality. Perspectives in Biology and Medicine 31:407-428.

Frisancho, R. (1974) Triceps Skinfold and Upper Arm Muscle Size Norms for Assesment of Nutritional Status. American Journal of Clinical Nutrition 27:1052-1058.

Frisancho, R. and P.N. Flegel (1982) Advanced Maturation Associated with Centripetal Fat Pattern. Human Biology 54:717-727.

Gajdusek, D.C. (1963) Motor-neuron Disease in Natives of New Guinea. New England Journal of Medicine 269:474-476.

Gajdusek, D.C. (1984) Environmental Factors Provoking Physiological Changes which Induce Motor Neuron Disease and Early Neuronal Aging in High Incidence Foci in the Western Pacific: Calcium Deficiency-induced Secondary Hyperparathyroidism and Resultant CNS Deposition of

Calcium and Other Metallic Cations as the Cause of ALS and PD in High Incidence Foci. In Progress in Motor Neuron Disease. F.C. Rose, ed. Kent: Pitman Books, Ltd.

Garn, S.M. (1975) Bone-loss and Aging. In The Physiology and Pathology of Aging, R. Goldman and M. Rockstein, eds. New York: Academic Press.

Garrey, P.J. and W.C. Hunt (1987) The Relative Merits of Dietary and Biochemical Information used to Assess Nutritional Status of the Elderly. American Journal of Physical Anthropology 72:201-202.

Garrey, P.J., J.S. Goodwin, W.C. Hunt, E.M. Hooper and A.G. Leonard (1982) Nutritional Status in a Healthy Elderly Population: Dietary and Supplemental Intakes. American Journal of Clinical Nutrition 36:319-331.

Garruto, R.M. (1989) Amyotrophic Lateral Sclerosis and Parkinsonism-dementia of Gaum: Clinical, epidemiological and genetic patterns. American Journal of Human Biology 1:367-382.

Garruto, R.M. and D.C. Gajdusek (1985) Factors Provoking the High Incidence of Amyotrophic Lateral Sclerosis and Parkinsonism-dementia of Guam: Deposition and Distribution of Toxic Metals and Essential Minerals in the Central Nervous System. In Normal Aging, Alzheimer's Disease and Senile Dementia: Aspects of Etiology, Pathogensis, Diagnosis, and Treatment. C.G. Gottfries, ed. Bruxelles: Editions de l'Universite'.

Garruto, R.M., C.C. Plato, N.C. Myrianthopoulos, M.S. Schanfield and D.C. Gajdusek (1983) Blood Groups, Immunoglobulin Allotypes and Dermatoglyphic Features of Patients with Amyotrophic Lateral Sclerosis and Parkinsonism-dementia of Guam. American Journal of Medical Genetics 14:289-298.

Garruto, R.M., R. Yanagihara, D.C. Gajdusek and D.M. Arion (1984) Concentrations of Heavy Metals and Essential Minerals in Garden Soil and Drinking Water in the Western Pacific. In Amytrophic Lateral Sclerosis in Asia and Oceania. K-M. Chen and Y. Yase, eds. Taipei: National Taiwan University.

Garruto, R. Yanagihara, R. and Gajdusek, D.C. (1985) Disappearence of High-incidence Amyotrophic Lateral Sclerosis and Parkinsonism-dementia on Guam. Neurology 35:193-198.

Gillum, R.F. (1987) The Association of Body Fat Distribution with Hypertension, Hypertensive Heart Disease, Coronary Heart Disease, Diabetes and Cardiovascular Risk Factors in Men and Women Aged 18-79 years. Journal of Chronic Diseases 40:421-428.

Goodwin, J.S. and P.J. Garrey (1988) Lack of Correlation Between Indices of Nutritional Status and Immunological Function in Elderly Humans. Journal of Gerontology 43:M46-M49.

Gould, K.G., Flint, M. and Graham, C.E. (1981) Chimpanzee Reproductive Senescence: A Possible Model for Evolution of the Menopause. Maturitas 3:157-166.

Graham, C.E., O.R. Kling and R.A. Steiner (1979) Reproductive Senescence in Female Nonhuman Primates. In Aging in Nonhuman Primates. D.M. Bowden, ed. New York: Van Nostrand Reinhold.

Hamilton, W.D. (1964) The Genetical Evolution of Social Behavior. Journal of Theoretical Biology 7:1-52.

Hamilton, W.D. (1966) The Moulding of Senescence by Natural Selection. Journal of Theoretical Biology 12:12-45.

Hardpending, H., A. Rogers and P. Draper (1987) Human Sociobiology. Yearbook of Physical Anthropology 30:127-150.

Heston, L.L. (1984) Down's Syndrome and Alzheimer's Dementia: Defining an Association. Psychiatr. Dev. 2:287-294.

Hoff C., R.M. Garruto, and N.M. Durham (1989) Human Adaptability and Medical Genetics. In Human Population Biology: A Transdisciplinary Science. M.A. Little and J.D. Hass, eds. Oxford: Oxford University Press, 69-81.

Hoffman, P.M., D.S. Robbins, C.J. Gibbs, Jr., and D.C. Gajdusek (1983) Immune Function Among Normal Guamanians of Different Age. Journal of Gerontology 38:414-419.

Hogden, G.D., A.L. Goodman, A. O'Conner and D.K. Johnson (1977) Menopause in Rhesus Monkeys: Model for Study of Disorders in Human Climacteric. American Journal of Obstetrics

and Gynecology 127:581-584.
Hoover, S.L. and J.S. Siegel (1986) International Demographic Trends and Perspectives on Aging. Journal of Cross-Cultural Gerontology. 1:5-30.
Howell, N. (1979) Demography of the Dobe !Kung. New York: Academic Press.
Howell, N. (1982) Village Composition Implied by a Paleodemographic Life Table: The Libben site. American Journal of Physical Anthropology 59:263-269.
Hrdy, S.B. (1981) "Neposits" and "Altruists": The Behavior of Old Females among Macaques and Langur Monkeys. In Other Ways of Growing Old: Anthropological Perspectives. P.T. Amoss and S. Harrell, eds. Stanford, CA: Stanford University Press.
İşcan, M.Y. (1988) Rise of Forensic Anthropology. Yearbook of Physical Anthropology 31:203-229.
İşcan, M.Y., S.R. Loth and R.K. Wright (1984a) Metamorphosis at the Sternal Rib End: A New Method to Estimate Age at Death in White Males. American Journal of Physical Anthropology 65:145-156.
İşcan, M.Y., S.R. Loth and R.K. Wright (1984b) Age Estimation from the Rib by Phase Analysis in White Males. Journal of Forensic Science 29:1094-1104.
Jackes, M.K. (1985) Pubic Symphysis Age Distributions. American Journal of Physical Anthropology 68:281-300.
Joachim, C.L. and D.J. Selkoe (1989) Minireview: Amyloid Protein in Alzheimer's Disease. Journal of Gerontology: Biological Sciences 44:B77-B82.
Johnson, T.E. (1988) Genetic Specification of Life Span: Processes, Problems, and Potentials. Journal of Gerontology: Biological Sciences 43:B87-B92.
Jones, P.R.M., P.S.W. Davies and N.G. Norgan (1986) Ultrasound Measurements of Subcutaneous Adipose Tissue Thickness in Man. American Journal of Physical Anthropology 71:359-363.
Jurmain, R. (1989) Trauma, Degenerative Disease, and Other Pathologies among the Gombe Chimpanzees. American Journal of Physical Anthropology 80:229-237.
Katona-Apte, J. (1984) Cross-cultural Aspects of Nutrition and Longevity. In Nutrition in Gerontology. J.M. Ordy, D. Hartman and R. Alfin-Slater, eds. New York: Raven Press.
Katz, D. and J.M. Suchey (1986) Age Determination of the Male Os Pubis. American Journal of Physical Anthropology 69:427-436.
Kerley, E.R. and D.H. Ubelaker (1978) Revisions in the Microscopic Method of Estimating Age at Death in Human Cortical Bone. American Journal of Physical Anthropology 49:545-546.
Kinner, B. and W. Ries (1986) Investigation of the Influence of Overnutrition and Dietetic Restriction on Health and Aging. Zeitschrift fur Alternsforsh 41:225-232.
Kirkwood, T.B.L. and R. Holiday (1986) Ageing as a Consequence of Natural Selection. In The Biology of Human Ageing. A.H. Bittles and K.J. Collins, eds. Cambridge: Cambridge University Press.
Koertvelyessy, T.A., M.H. Crawford and J. Hutchinson (1982) PTC Tasting Threshold Distributions and Age in Mennonite Populations. Human Biology 54:635-646.
Kohrs, M.B. and D.M. Czajka-Narins. (1986) Assessing the Nutritional Status of the Elderly. In Contemporary Issues in Clinical Nutrition, Aging, and Health. E.A. Young, ed. New York: Alan R. Liss.
Lamb, M.J. (1986) Insects as Models for Testing Theories of Aging. In The Biology of Human Ageing. A.H. Bittles and K.J. Collins, eds. Cambridge: Cambridge University Press.
Lancaster, J.B. and King, B.J. (1985) An Evolutionary Perspective on Menopause. In In Her Prime: A New View of Middle-Aged Women. J.K. Brown and V. Kerns, eds. South Hadley, MA: Bergin and Garvey.
Läpidus L., C. Bengtsson, B. Lärsson, K. Pennert, E. Rybo and L. Sjostrom (1984) Distribution of Adipose Tissue and Risk of Cardiovascular Disease and Death: A 12 Year Follow-up of Participants in the Population Study of Women in Gothenburg, Sweden. British Medical Journal 289:1257-1261.

Lärsson B., K. Svärdsudd, L. Welin, L. Wilhelmsen, P. Björntorp and G. Tibblin (1984) Abdominal Adipose Tissue Distribution, Obesity and Risk of Cardiovascular Disease and Death: A 13 Year Follow-up of Participants in the Study of Men Born in 1913. British Medical Journal 288:1401-1404.

Loth, S.R. and M.Y. Işcan (1988) Skeletal Aging Techniques: The Good, the Bad, and the Equivocal. American Journal of Physical Anthropology 75:241.

Makinodan, T. and E. Yunis (1977) Immunology and Aging. New York: Plenum Press.

Mann, D.M. (1985) The Neuropathology of Alzheimer's Disease: A Review with Pathogenetic, Aetiological and Therapeutic Considerations. Mechanisms of Ageing and Development 31:213-255.

Manton, K.G. (1986a) Cause Specific Mortality Patterns Among the Oldest Old: Multiple Cause of Death Trends 1968 to 1980. Journal of Gerontology 41:282-289.

Manton, K.G. (1986b) Past and Future Life Expectancy Increases at Later Ages: Their Implications for the Linkage of Chronic Morbidity, Disability, and Mortality. Journal of Gerontology 41:672-681.

Manton, K.G. and E. Stallard (1984) Recent Trends in Mortality Analysis. Orlando: Academic Press.

Manton, K.G., S.S. Poss and S. Wing (1979) The Black/White Mortality Crossover: Investigation from the Perspective of the Components of Aging. The Gerontologist 19:291-300.

Maples, W.R. (1978) An Improved Technique using Dental Histology for Estimation of Adult Age. Journal of Forensic Science 23:747-770.

Markides, K.S. and R. Machalek (1984) Selective Survival, Aging and Society. Archives of Gerontology and Geriatrics, 3:207-222.

Martin, G. (1987) Interactions of Aging and Environmental Agents: The Toxicological Perspective. In Environmetal Toxicity and the Aging Process. S.R. Baker and M. Rogul, eds. New York: Alan R. Liss.

Masoro, E.J. (1987) Biology of Aging: Current State of Knowledge. Archives of Internal Medicine 147:166-169.

Masoro, E.J. (1988) Minireview: Food Restriction in Rodents: An Evaluation of its Role in the Study of Aging. Journal of Gerontology 43:B59-B64.

Masoro, E.J. (1989) Food Restriction Research: Its Significance for Human Aging. American Journal of Human Biology 1:339-345.

Masoro, E.J., M.S. Katz and C.A. McMahon (1989) Evidence for the Glycation Hypothesis of Aging from the Food-restricted Rodent Model. Journal of Gerontology: Biological Sciences 44:B20-B22.

Mayer, P.J. (1982) Evolutionary Advantage of the Menopause. Human Ecology 10:477-494.

Mayer, P.J. (1987) Biological Theories of Aging. In The Elderly as Modern Pioneers. P. Silverman, ed. Bloomington, IN:Indiana University Press.

Mayer, P.J., M.O. Bradley and W.W. Nichols (1986a) No Change in DNA Damage or Repair of Single- or Double-strand Breaks as Human Diploid Fibroblasts Age in Vitro. Experimental Cell Research 166:497-509.

Mayer, P.J., M.O. Bradley and W.W. Nichols (1986b) Incomplete Rejoining of DNA Double-strand Breaks in Unstimulated Normal Human Lymphocytes. Mutation Research 166:275-285.

Mayer, P.J., M.O. Bradley and W.W. Nichols (1987) The Effect of Mild Hypothermia (34C) and Mild Hyperthermia (39C) on DNA Damage, Repair and Aging of Human Diploid Fibroblasts. Mechanisms of Ageing and Development 39:203-222.

Mazess, R.B. (1982) On Aging Bone Loss. Clinical Orthopedics and Related Research 165:239-252.

Mazess, R.B. and R.W. Mathisen (1982) Lack of Unusual Longevity in Vilcabamba, Ecuador. Human Biology 54:517-524.

McCay, C.M., M.F. Crowell and L.A. Maynard (1935) The Effect of Retarded Growth on the Length of Life Span and upon Ultimate Body Size. Journal of Nutrition 10:255-263.

Meier, R.J., C.S. Goodson and E. Roche (1986) Dermatoglyphic Development and Rate of

Maturation. American Journal of Physical Anthropology 69:239.
Memeo, S.A., L. Piantanelli, G. Mazzufferi, L. Guerra, M. Nikolitz and N. Fabris (1982) Age Related Patterns of Immunoglobulin Serum Levels in the Quechua Indians of the Andean Mountains. International Journal of Biometeorology 26:49-52.
Merry, B.J. (1986) Dietary Manipulation of Ageing: An Animal Model. In The Biology of Human Ageing. A.H. Bittles and K.J. Collins, eds. Cambridge: Cambridge University Press.
Micozzi, M. (1987) Cross-Cultural Correlations of Childhood Growth and Adult Breast Cancer. American Journal of Physical Anthropology 73:525-537.
Molleson, T.I. (1981) The Archaeology and Anthropology of Death: What the Bones Tell Us. In Mortality and Immortality, S.H. Humphries, ed. London: Academic Press.
Molleson, T.I. (1986) Skeletal Age and Paleodemography. In The Biology of Human Ageing. A.H. Bittles and K.J. Collins, eds. Cambridge: Cambridge University Press.
Moment, G.B., ed. (1982) Nutritional Approaches to Aging Research. Boca Raton, FL: CRC Press Inc.
Moore, M.J. (1981) Menarch, Menopause and Aging. American Journal of Physical Anthropology 54:256.
Moore, M.J. (1987) The Human Life Span. In The Elderly as Modern Pioneers. P. Silverman, ed. Bloomington, IN:Indiana University Press.
Mueller, W.H. (1982) The Changes with Age of the Anatomical Distribution of Fat. Social Science and Medicine 16:191-196.
Mueller, W.H., M.I. Deutsh, R.M. Malina, D.A. Bailey and R.L. Mirwald (1986) Subcutaneous Fat Topography: Age Changes and Relationships to Cardiovascular Fitness in Canadians. Human Biology 58:955-973.
Nam, C.B., N. Weatherby and K. Ockay (1978) Causes of Death which Contribute to the Mortality Crossover Effect. Social Biology Winter 25:306-314.
Ndaba, N. and S.J.D. O'Keefe (1986) The Nutritional Status of Black Adults in Rural Districts of Natal and Kwazulu South Africa. South African Medical Journal 68:569-590.
Newell-Morris, L., V. Moceri and W. Fujimoto (1989) Gynoid and Android Fat Patterning in Japanese-American Men: Body Build and Glucose Metabolism. American Journal of Human Biology 1:73-86.
Norgan, N.G. (1987) Fat Patterning in Papua New Guineans: Effects of Age, Sex and Acculturation. American Journal of Physical Anthropology 74:385-392.
Ordy, J.M., D. Hartman and R. Alfin-Slater (1984) Aging, Volume 26: Nutrition in Gerontology, New York: Raven Press.
Palmore, E.B. (1982) Predictors of the Longevity Difference: A 25-year Follow-up. The Gerontologist 22:513-518.
Pearl, R.D. (1931) Studies on Human Longevity III: The Inheritance of Longevity. Human Biology 3:245-253.
Pearl, R. and H.D. Pearl (1934) The Ancestry of the Long-Lived. Baltimore: Johns Hopkins University Press.
Pearson, J.D. and D.E. Crews (1989) Evolutionary, Biosocial, and Cross-Cultural Perspectives on the Variability in Human Biological Aging. American Journal of Human Biology 1:303-306.
Pearson, M.B., E.J. Bassey and M.J. Bendall (1985) Muscle Strength and Anthropometric Indices in Elderly Men and Women. Age and Ageing 14:49-54.
Plato, C.C. (1987) The Effects of Aging on Bioanthropological Variables: Changes in Bone Mineral Density with Increasing Age. Colloquium in Anthropology 11:59-72.
Reaven, G.M. and E.P. Reaven (1985) Age, Glucose Intolerance, and Non-insulin-dependent Diabetes Mellitus. Journal of the Geriatric Society 33:286-290.
Rivlin, R.S. and E.A. Young eds. (1982) Symposium on Evidence Relating Selected Vitamins and Minerals to Health and Disease in the Elderly Population in the United States. American Journal of Clinical Nutrition 36:977-1086.
Roche, A.F. (1966) Aging in the Human Skeleton. The Medical Journal of Australia 2:943-946.

Rogers, L.A. and M.H. Crawford (1981) Population Genetics of Aging and Longevity among Mennonites of Kansas, USA. American Journal of Physical Anthropology 54:269.

Rose, M.R. and J.L. Graves, Jr. (1989) Minireview: What Evolutionary Biology can do for Gerontology. Journal of Gerontology: Biological Sciences 44:B27-B29.

Roth, G.S., R.G. Cutler and D.K. Ingram (1988) Effect of Caloric Restriction on Primate Aging Rate. The Gerontologist 28:223A.

Rundgren, A., S. Eklund and R. Jonson (1984) Bone Mineral Content in 70-and 75-year-old Men and Women: An Analysis of Some Anthropometric Background Factors. Age and Ageing 13:6-13.

Ryan, A.S., G.A. Martinez, J.L. Wysong and M.A. Davis (1989) Dietary Patterns of Older Adults in the United States, NHANES II (1976-1980). American Journal of Human Biology 1:321-330.

Sattenspiel, L. and H. Hardpending (1983) Stable Populations and Skeletal Age. American Antiquity 48:489-498.

Schweber, M. (1985) A Possible Unitary Genetic Hypothesis for Alzheimers Disease and Downs Syndrome. Annals of the New York Academy of Sciences 450:223-238.

Seidell, J.C., A. Oosterlee, M.A. Thijssen, J. Burema, P. Deurenberg, J.G. Hautvast and J.H. Ruijs (1987) Assessment of Intra-abdominal and Subcutaneous Abdominal Fat: Relation between Anthropometry and Computed Tomography. American Journal of Clinical Nutrition 45:7-13.

Shepard, R.J., P.R. Kofsky, J.E. Harrison, K.G. McNeil and A.G. Krondel (1985) Body Composition of Older Female Subjects: New Approaches and their Limitations. Human Biology 57:671-686.

Short, R., D.D. Williams, and D.M. Bowden (1987) Cross-sectional Evaluation of Potential Biological Markers of Aging in Pigtailed Macaques: Effects of Age, Sex, and Diet. Journal of Gerontology 42:644-654.

Siu, M.L., M. Mazariegos, L. Luhfeld, N.W. Solomons and O. Pineda (1987) Nutritional Status, Body Composition and Estimated Weight Loss in Institutionalized Guatemalan Elderly. Federation Proceedings 46:900.

Sidhu, L.S., P. Singal and D.K. Kansal (1983) Processes of Maturation and Senescence in Two Communities of Punjab (India) as Studied from Some Morphological Measures. Anthropologischer Anzeiger 41:149-159.

Solomon, L. (1979) Bone Density in Aging Caucasian and African Populations. Lancet 1326-1329.

Sørensen, T.I.A., G.G. Nielsen, P.K. Andersen, and T.W. Teasdale (1988) Genetic and Environmental Influences on Premature Death in Adult Adoptees. New England Journal of Medicine 318:727-732.

Stini, W.S., R.J. Harrington, S.C. McCombie (1983) Bone Mineral Loss in an Aging Population in an Affluent Retirement Community. American Journal of Physical Anthropology 60:257-258.

Stini, W.S. (1984) The Significance of Body Composition in Human Aging. Journal of the Indian Anthropological Society 19:80-91.

Stini, W.S. (1987) Bone Mineral Loss and Stature Decrease among Retirees of Differing Levels of Affluence. American Journal of Physical Anthropology 72:258.

Stout, R.W. and V. Crawford (1988) Active Life Expectancy and Terminal Dependency: Trends in Long-term Geriatric Care over 33 Years. Lancet, 1:281-283.

Suchey, J.M. and S. Brooks (1987) Use of the Suchey-Brooks System for Aging the Male Os Pubis. American Journal of Physical Anthropology 72:259.

Suzman, R. and M.W. Riley (1985) Introducing the "Oldest Old". Millbank Memorial Fund Quarterly 63:177-186.

Swedlund, A.C., R.S. Meindl, J. Naylor and M.I. Gradie (1983) Family Patterns in Longevity and Longevity Patterns of the Family. Human Biology 55:115-129.

Szathmary, E.J.E. and N. Holt (1983) Hyperglycemia in Dogrib Indians of the Northwest Territories, Canada: Association with Age and a Centripetal Distribution of Body Fat. Human

Biology 55:493-515.
Tengerdy, R.P. (1980) Disease Resistance: Immune Response. In Vitamin E: A Comprehensive Treatise. L.J. Machlin, ed. New York: Marcel Dekker.
Timeras, P.S. (1988) Physiological Basis of Aging and Geriatrics. New York: Macmillan.
Tobin, J.D. (1987) Nutritional Impact of Physiological Variables in the Baltimore Longitudinal Study of Aging (BLSA). American Journal of Physical Anthropology 72:263.
Torrey, B.B., K.G. Kinsella and C.M. Taeuber (1987) An Aging World (Advance Report). US Department of Commerce. Bureau of the Census.
Trinkhaus, E. and D.D. Thompson (1987) Femoral Diaphyseal Histomorphometric Age Determinations for the Shanidar 3, 4, 5, and 6 Neanderthals and Neanderthal Longevity. American Journal of Physical Anthropology 72:123-129.
Turnquist, J.E. (1983) Caging and Joint Mobility. American Journal of Physical Anthropology 60:263.
Vaupel, J.W. (1988) Inherited Frailty and Longevity. Demography 25:277-287.
Vague, J. (1956) The Degree of Masculine Differentiation of Obesities: A Factor Determining Predisposition to Diabetes, Atherosclerosis, Gout, and Uric Acid Calculous Disease. American Journal of Clinical Nutrition 4:20-34.
Vague, J., P. Björntorp, B. Guy-Grand, M. Rebuffe-Scrive and P. Vague (1985) Metabolic Complications of Human Obesities. Amsterdam: Excerpta Medica.
Vir, S.C. and A.H. Love (1980) Anthropometric Measurements in the Elderly. Gerontology 26:1-8.
Walford, R.L., R.K. Liu, M. Gerbase-Delima, M. Mathies and G.S.Smith (1974) Longterm Dietary Restriction and Immune Function in Mice. Mechanisms of Ageing and Development 2:447-454.
Walford, R.L. (1983) Maximum Life Span. New York: W.W. Norton.
Walford, R.L. (1986) The 120-Year Diet. New York: Simon and Schuster.
Walker, P.L., J.R. Johnson and P.M. Lambert (1988) Age and Sex Biases in the Preservation of Human Skeletal Remains. American Journal of Physical Anthropology 76:183-188.
Warner, H.R., R.N. Butler, R.L. Sprott and E.L. Schneider (1987) Aging, Volume 31: Modern Biological Theories of Aging. New York: Raven Press.
Weindruch, R. (1984) Dietary Restriction and the Aging Process. In Free Radicals in Molecular Biology, Aging and Disease. D. Armstrong, R.S. Sohol, R.G. Culter, and T.F. Slater, eds. New York: Raven Press.
Weindruch, R., P.H. Naylor, A.L. Goldstein and R.L. Walford (1988) Influences of Aging and Dietary Restriction on Serum Thymosin Alpha-one Levels in Mice. Journal of Gerontology: Biological Sciences 43:B40-B42.
Weiss, K.M. (1981) Evolutionary Perspectives on Human Aging. In Other Ways of Growing Old: Anthropological Perspectives. P.T. Amoss and S. Harrell, eds. Stanford, CA: Stanford University Press.
Weiss, K.M. (1984) On the Number of Members of Genus Homo who Have Ever Lived, and Some Evolutionary Interpretations. Human Biology 56:637-649.
Weiss, K.M. (1989) Are Known Chronic Diseases Related to the Human Lifespan and its Evolution? American Journal of Human Biology 1:307-319.
Williams, D.D. and D.M. Bowden (1984) A Nonhuman Primate Model for the Osteopenia of Aging. In Comparative Pathobiology of Major Age-Related Diseases. D.G. Scarpelli and G. Migaki, eds. New York: Alan R. Liss.
Williams, G.C. (1957) Pleiotropy, Natural Selection, and the Evolution of Senescence. Evolution 11:398-411.
Wing, S., K.G. Manton, E. Stallard, C. Hames and H.A. Tyroler (1985) The Black/White Mortality Crossover: An Investigation in a Community-based Cohort. Journal of Gerontology 40:78-84.
Workshop of European Anthropologists (1980) Recommendations for Age and Sex Diagnoses of Skeleton. Journal of Human Evolution 9:517-549.

Wright, A.F. and L.J. Whalley (1984) Genetics, Ageing and Dementia. British Journal of Psychiatry 145:20-38.

Wurtman, J.J., H. Liberman, R. Tsay, T. Nader and B. Chein (1988) Caloric and Nutrient Intakes of Elderly and Young Subjects Measured under Identical Conditions. Journal of Gerontology: Biological Sciences 43:B174-B180.

Yanagihara, R., R.M. Garruto, D.C. Gajdusek, A. Tomita, T. Uchikawa, Y. Konagaya, K.M. Chen, I. Sobue, C.C. Plato and C.J. Gibbs (1984) Calcium and Vitamin D Metabolism in Guamanian Chamorros with Amyotrophic Lateral Sclerosis and Parkinsonism-dementia. Annals of Neurology 15:42-48.

Young, V.R. and N.S. Scrimshaw (1979) Genetic and Biological Variability in Human Nutrient Requirements. American Journal of Clinical Nutrition 32:486-500.

J. NEIL HENDERSON

## 2. ANTHROPOLOGY, HEALTH AND AGING

INTRODUCTION

Most people wish for a long and healthy life. In fact Leo Simmons (1945), in his pioneering article on cross-cultural aging, concluded that the wish for long life while remaining active was a human universal. Cowgill and Holmes (1972) found the same universal in a worldwide sample, specifying that two interacting co-variables are required to achieve a desirable long life: longevity coupled with the ability to function well in everyday life.

James Fries (1980, 1984) suggests that the goal of a long life with high functional status necessitates the compression of late life morbidity into the shortest time span possible. Rather than a long life with years of chronic disease and progressive functional decline, the older adult would ideally maintain a plateau of reasonable functional ability until a final few months or so of morbidity leading to death. Much of the human potential for maximizing life span has been accomplished, and so this ideal configuration has been partially achieved. Humans have a life span potential of about 115 years (Kirkwood 1985) and are now living almost to age 80 (National Center on Health Statistics 1985). However, the actual compression of morbidity into a small package of time fails us as reflected in the increase of chronic, incurable disease. It is as if the elusive fountain of youth has been changed into a dreaded "fountain of longevity."

Health and aging must also be understood in the context of how recent "old age" is among humans. The first Homo sapiens lived in the middle paleolithic, about 100,000 years ago. Skeletal remains show that most of these Neandertal people died at about 30 years of age (Moore 1987). Only a few skeletons indicate an age of death in the 60th decade. In the 1990's, the average human life span will be between 75-80 years of age. Today, a person who survives to age 60 can expect to live another 20.3 years. Only two decades ago, Americans would plan for their funeral arrangements once they attained the ripe old age of 60, while now, arrival at age 60 is the beginning of an additional two decades. Even a person living to age 85 can expect to live another 6.2 years (Monthly Vital Statistics Report 1983).

This discussion of health and aging will use "health" in the broad anthropological sense to include biocultural features of physical and mental health. Also included will be indirect factors such as the effects of social networks and the health care system in which older adults find themselves (e.g., nursing homes, acute care hospitals, day care centers, and single room occupancy hotels, age-segregated villages, etc.). A detailed discussion of medical anthropology perspectives and their use in age-related studies appears below.

## UNIQUE ASPECTS OF GERIATRIC HEALTH

Examining health in late life from an anthropological perspective requires some understanding of physiological peculiarities that may be experienced by older adults. As Fabrega (1974:41) suggests, "Given a concern with behavior and its organization and a desire to probe medical issues in an incisive manner, an awareness of the biological components of disease would appear mandatory." In addition to the biological components, other questions to be addressed include: How will the older adult interpret and react to symptoms? How will health care practitioners relate to them as "geriatric patients"? What is different about geriatric medicine compared to ordinary adult internal medicine? The answers to these questions spark ideas for anthropological research in health and aging.

Unique aspects of geriatric health are succinctly captured in a list presented to medical students regarding diagnostic treatment approaches to geriatric patients. The geriatric profile often consists of five telling characteristics commonly experienced by older patients (Ham 1983; also see Besdine 1982). First, the geriatric patient is more apt to come to a clinician with atypical symptoms for a given condition. People of all ages are subject to experiencing atypical symptoms, but it is considered more frequent among the elderly patient. An example is the case of an 82-year old female who was brought to the emergency room by a family member who reported that the patient had complained of nausea, general malaise, and loss of appetite during the previous eight hours. This patient was at the moment of the clinical history-taking experiencing a serious myocardial infarction. However, the classic presenting symptoms of "crushing sub-sternal pain radiating into the left arm" was missing. Similarly, the geriatric patient may not have a significantly elevated temperature in the presence of severe infections. The geriatric practitioner must, therefore, be vigilant in considering many more alternative causes for symptoms in the geriatric patient.

Second, older adults commonly have several active, diagnosable diseases at any given point in time. The existence of simultaneous multiple pathologies does not necessarily mean that the patient is experiencing severe functional loss (Brody and Foley 1985). It may be that most of these diseases are of a chronic but not totally incapacitating type. However, the middle aged adult usually encounters the health care practitioner with one primary complaint which is evaluated and treated. The health care practitioner evaluating the geriatric patient must generate the diagnostic roster for a given patient and then decide which condition to treat most aggressively and without interference to other conditions or therapeutic strategies.

Third, the older adult has a generally reduced ability to "bounce back" from physical and even emotional trauma compared to young and middle-aged adults. The reduced homeostasis of older persons causes them to be at higher risk of experiencing relatively minor afflictions in major ways. Also, the reduced efficiency of regaining homeostasis may be mistakenly interpreted by practitioners and patients as the onset of the "expected breakdowns" of old age.

Fourth, a related aspect of reduced homeostasis is the inefficient metabolism of medications either self-prescribed or physician-prescribed. Standard dosages are often considered by geriatric practitioners to be too high for the older patient, particularly if the medication is for long-term use (Rowe and Besdine 1982). The geriatric patient's inefficiency in metabolizing medications combined with a standard schedule for doses, causes additional dosages to be taken before the previous dosage has been completely used by the body. The result is a toxic build-up of medications. It is not uncommon for the geriatric practitioner to occasionally give the patient a "drug holiday". Also, a new geriatric patient may be told to discontinue or drastically reduce all medications with the observation that the patient, at least temporarily, improves.

An additional problem with medications that requires the geriatric practitioner's attention is the likelihood of a patient seeing several different practitioners for the several different ailments that they are experiencing. The medications prescribed by each practitioner are unknown to the other practitioners. The patient may go home and dutifully take all the prescriptions according to the labels and may experience unnecessary worsening of their condition because of the polypharmacy phenomenon.

Lastly, geriatric specialists consider the older patient to be at high risk of iatrogenic disease. This type disease is the result of deleterious outcomes of therapeutic interventions. Etymologically, iatrogenic disease would implicate only the physician as the source of the unintended noxious effects of treatments. However, health care practitioners in general and the health care system at large often interact as the source of low quality health care experiences for the older adult.

## MEDICAL ANTHROPOLOGY AND STUDIES OF AGING

Due to the diverse nature of anthropological research and aging, a single broad category cannot capture the variant types of studies. Consequently, research in health and aging from an anthropological perspective can be examined in a four-part scheme which includes 1) medical anthropology and aging, 2) general community studies, 3) age-segregated community studies, and 4) disease-specific studies. The bulk of the anthropological literature on health and aging can be placed within each one of these four categories. While the four categories of research are not, fully representative of the universe of all perspectives on the triplet of anthropology, health, and aging, they are purposely conceived at a high level of abstraction in order to provide a framework based on the actual history of the research.

The section on medical anthropology and aging will review some of the earliest anthropological research in aging and health care in clinical and community settings. However, the general focus here is on research which is pioneering and prospective in its impact on the field. The second section will examine aging research conducted in general communities ranging from ordinary residential living patterns to public housing and single room occupancy hotels. Within this section research on those more invisible elderly embedded in the community but who are experiencing the effects of social isolation is also discussed. The third section will examine age-segregated community studies from two perspectives: a) research conducted in non-institutional settings which are used by older adults, and b) institutional settings in which ethnographic research has been conducted. The fourth section will examine research on two specific health care problems commonly experienced in late life: dementia and cerebrovascular stroke.

*Medical Anthropology and Aging*

Health is conceptualized by anthropologists as enmeshed in biological, psychosocial, and sociocultural contexts. Because physical symptoms are brought to conscious and interpreted in the context of a culturally based meaning and value system, all of health and disease is fundamentally a social science. In other words, "Disease and illness exist, then, as constructs in particular configurations of social reality" (Kleinman 1980:73). Useful histories of the development of medical anthropology concepts can be found in several sources (Polgar 1962; Scotch 1963; Alland 1966; Foster 1975; Wellin 1978; Foster and Anderson 1978; Brenner, Mooney and Nagy 1980; Eisenberg and Kleinman 1981).

Medical anthropologists have developed several conceptual models for the study of health and disease which represent the comprehensive nature of the human condition when life is experienced under "altered conditions" (Virchow 1849). For example, Fabrega (1974) analyzes disease in terms of "behavioral frameworks" which include behaviors like task-action, role enactment, and other culturally influenced reactions and interpretations of symptoms. Antonovsky (1979) developed the "salutogenic model of health" as a way of explaining health more than pathology. Healthy people have a highly developed sense of confidence, called the "sense of coherence" combined with "generalized resistance resources" which include mainly psychosocial phenomena such as knowledge, coping ability, social supports, cultural stability, and include genetically derived resistance resources.

Additional conceptual models for understanding health and disease include Engel's "biopsychosocial model" (Engel 1977, 1980). The term "biopsychosocial" (cumbersome as it is) does not convey the entire model that includes reference to the components of community, culture, subculture, and society that are present in Engel's discussion albeit relatively weakly developed.

Nearly concurrent with Engel's work was a paper in the *Annals of Internal Medicine* (1978) by Kleinmen, Eisenberg, and Good. The clinic was given a recipe for quickly (if somewhat superficially) determining how patients conceived of their illnesses. The patient's "explanatory model" (further elucidated in Kleinman 1980) was elicited by questions which probe for the patient's beliefs about their illness and what the patient thought was the cause and should be the cure. This emphasis on the patient (i.e., native) is reminiscent of the ethnomedical model which examines emic health systems, beliefs and behaviors (Fabrega 1974, 1975; Hughes 1978). Also, further work on the clinical encounter has been done by Hill, Fortenberry, and Stein (in press) in which they suggest a cautious approach to the clinical use of practitioner-determined cultural variables. Their concern is that superficially derived and understood cultural variables may essentially be stereotypes of dubious accuracy and limited use.

Other social scientists have included broad sociocultural elements in studies of disease such as the effect of the social environment on physical health (i.e., social epidemiology) (Berkman 1980) and the effects of social networks as coping resources to aid prevention or recovery from illness episodes (McKinlay 1980).

It is a truism to gerontologists that chronological age alone is the most dispensable variable of all when trying to explain some human phenomenon. Older adults persist in active interaction with self, lay, and professional health care systems (Haug 1981). Therefore, the above cited health models are quite germane to older adults and the analysis of their health or disease circumstances.

Medical anthropology and aging research combines health and disease analysis within the biocultural context of the late life experience as experienced in a given society, time, and place. The reader will note the anthropological (perhaps Darwinian) theme of adaptation to various and changing sociocultural environments. Medical anthropologists, from a biocultural perspective, add to the "sociocultural environment" the dimension of adaptation to changing external environments with a body experiencing changing internal environments that cannot be pre-experienced. The life-long capacity for adaptation is truly remarkable.

Medical anthropologist Otto von Mering can be credited with some of the earliest investigations of aging and well-being. For example in 1957 von Mering reported his ethnographic study of a geriatric ward in a mental hospital in which prevailing cultural de-valuing of old age led to a general therapeutic breach leaving the geriatric patient bereft of psychosocial care. Yet, von Mering demonstrates that under the influence of a "remotivation program", the geriatric patient became more socially responsive and likely to be discharged to less confining types of care.

In "Cultural Values of Normal Senescence, Illness and Death: An Essay in Comparative Gerontology" (1958), von Mering developed a cross-cultural perspective on health and aging outside institutional walls. This article was published in a psychiatry journal. The cross-cultural element gave the reader a way to better conceptualize values about aging in their own society, self and practice.

Von Mering also contributed to the now classic *Handbook of Aging and the Individual*, a chapter titled "Sociocultural Background of the Aging Individual" (von Mering and Wenieger 1959). Von Mering and Wenieger took an adaptive approach to aging by calling for understanding of the unique and variant ways of experiencing late life in the presence of normal age-related changes, as well as those that may be pathological. That is, the analyst of aging and health must understand the "man-environment context of the complaint" (von Mering and Wenieger 1959). Later, Earley and von Mering (1969) described an outpatient clinic used by people who had suffered from a diffuse health aberration for years that had vexed the health care practitioners to the point of considering these patients malingerers. Moreover, several of the patients had been seen at the outpatient clinic for as long as eighteen years. By the time these patients had become "geriatric", they were suffering the effects of negative labeling on the part of the staff for apparent malingering but with the new additional negative status of "aged." However, Earley and von Mering (1969) consider this response to late life and ill health as an adaptive one for some of their elderly clinic population, because it allowed them to integrate the health problems of their early and middle adult years with the physical changes common to late life.

Still working in a hospital setting, von Mering (1969) observed "retirement from life into active ill health" as an all-to-common response to aging. Von Mering emphasizes the necessity of understanding the "biomedical matrix of aging." Specific concern for understanding the changing functional health status of late life as it interacts with chronic disease is a biomedical foundation within which sociocultural beliefs and behaviors must be considered. Twenty years ago, in a remarkable conceptual leap anticipating the present, von Mering (1969) predicted that "... careful scrutiny of the existing caretaking patterns is sometimes more relevant to treatment than is the diagnosis itself." Today, geriatricians' interest in functional status is as important as diagnosis and sometimes more so (Ham 1983). In the same paper, von Mering (1969) accented the importance of understanding the patient's cognitive map of their current health problems as an important management tool for the practitioner. Today, this is the essence of the Kleinman (1980) "explanatory model".

In 1973, Margaret Clark identified the study of aging from a cultural anthropological perspective as a relatively recent endeavor. Clark notes that (the famous 1945 Leo Simmons article notwithstanding) anthropologists who have done general ethnographies or who have specifically studied life-span phenomena have incorporated the elderly as subjects in studies of non-Western societies, albeit not usually as the focus of the study. Until the Clark article, studies on aging were perceived to be predominantly from sociology and psychology. By citing a number of contemporary studies on aging done by anthropologists Clark corrected this bias (e.g., Arth 1968; Holmes 1972; Munsel 1972; Adams 1972; Clark and Mendelson 1969; LeVine 1965; Plath 1972; von Mering 1969; Cowgill and Holmes 1972; Kiefer 1971; Ross 1972; Press and McKool 1972).

Clark (1973:80) further identified six salient aspects of aging which emerged during the twelve years prior to her article from the research of cultural anthropologists. These are: dying, decrement and disengagement, disease, dependency and regression, minority group status, and (life span) development. Clark stressed that these biocultural life events are universally shaped by culture and the cultural content of aging must be delineated in order to understand it. Furthermore, the cultural content regarding aging contains values and behaviors relevant to health and well-being which will promote or inhibit successful aging. For example, the primary American values of being independent and fully functioning effect health and self-esteem by influencing definitions of proper functioning and equate high health status with independence.

Clark concluded the article by reference to five adaptive tasks required for successful aging. These five adaptive tasks were brought forward from her research among community dwelling elderly in San Francisco (Clark and Anderson 1967). These researchers found that each task is required for successful reactions to the inevitable changes of aging (these will be explicated below). Finally, Clark concluded that anthropology offers the opportunity for a fine grained analysis via ethnography that is required to understand the adaptive reactions made by old people and to reveal them as vibrant and vital participants in their cultural arena.

George Foster and Barbara Anderson (1978) presented the life span perspective from an anthropological viewpoint in one of the first books using the appellation Medical Anthropology. In a section on life-span bioethics they showed that the "relative worthlessness of late life" in comparison to earlier stages of life is a recent cultural phenomenon in American society.

Significantly, Foster and Anderson (1978) summarize the American aging experience in terms of two basic models of coping: the "confront-the-system" model and the "escape-the-system" model. The confront-the-system model requires the active adaptation on the part of the older member of society. They cite the example of the Clark and Anderson research (1967) in which successful aging took place when older individuals were willing to understand their place within the system and carve a specific adaptive niche in it. The escape-the-system model is exemplified by the development and proliferation of age-segregated communities, including life care communities and complete but separate villages of old people marketed as retirement towns (Byrne 1974; Jacobs 1974).

Andrea Sankar (1984) provides a history of the Western medical construction of old age as disease and its diagnosis and symptoms. One problem with aging and Western medicine, she finds, is that "old" can be used as a diagnostic taxon. Thus, Sankar cites an example of a conversation between a physician and a pharmacist in which the physician "diagnosed" everyone over 95 years of age as "suffering old age." But, the pharmacist noted the case of a patient who was being treated specifically for a heart condition. The physician who was the medical director of a nursing home said that it was well established that by the time someone is over age 95, most of what is wrong with them is very likely due to their advanced age. The pharmacist then asked what diagnosis would be given to a 95 year old who is very healthy; the physician was nonplused and unable to continue the discussion.

Sankar also reports that the distinction between an actual disease and those symptoms that are related to old age (such as normal age related changes) are easily distinguished by the patient themselves while remaining elusive to their health care providers. In a very perceptive account of aging in the context of the family, Sankar points out that younger family members cannot pre-experience the exigencies of old age nor the physical changes and symptoms that may accompany old age. The older adult then enters a life experience which other members of their family cannot well understand.

The family must also experience their kin in late life as interacting with a health care system which is unable to effectively treat their chronic diseases. Consequently, a problem ensues in which the health care practitioner recoils from treating an elderly person for whom they perceive very little prospect of therapeutic outcome.

By way of contrast, Sankar (1984) also provides a window into the Chinese medical system and its perception of old age. Among the Chinese of Hong Kong, old age is not considered a disease, yet an association between advanced age and increasing physical complaints is made and accepted. Sankar also reports that her subjects paid more attention to their overall health status and function than to disease. Moreover, this was true of informants who were involved both in formal religious/healing activities as well as lay informants.

As in the United States, the Chinese elders seek treatment from healers mainly in regard to chronic illness. Although the elderly Chinese are very interested in health and maintaining health, they are much more prone to first engage in self-treatment of disease and reserve only those symptoms considered extremely serious for the review of Western trained physicians. Western-trained physicians were the least used health care practitioners observed during this study. Chinese trained physicians were most common followed by indigenous folk healers.

The Chinese family is a haven for the elderly person due in part to a Confucian ethic which venerates the elderly population. However, simple advanced age is not sufficient to reap the benefit of high status, support, and assistance. The older person must function in a meaningful way in the social context of the family so that care, support, and respect will be accorded them.

*General Community Studies*

In general community studies and those studies of age segregated communities to be considered below, an important view of health that emerges is the context of adaptation by older individuals to the communities and setting that they inhabit. First, elders in these communities share in or are subject to general cultural values, such as those described by Clark and Anderson in their seminal work, that have been adapted to their particular community. Second, elders of various health statuses and levels of functioning may creatively use strategies of values management and redefinition in order to adapt their health status and other limitations or capabilities to particular settings. Thus, the view taken here is that health is a socially complex phenomenon that is best seen in the context of the overall process of group and individual adaptation.

The Langley Porter Institute Studies in Aging initiated a program of gerontological research in 1959. The treatise by Margaret Clark and Barbara Anderson (1967) is one of the major products of this program which was funded in part by the National Institute of Mental Health and the State of California (Department of Mental Hygiene). Clark and Anderson discuss the analytical perspective of anthropology as part of the study of aging with a very clear interest in adaptive strategies used in late life by those who successfully age. Also, the mental health status of their subjects is a theme which runs throughout their work somewhat to the exclusion of the physical health of the subject (except by self report). The seminal project was undertaken in San Francisco with a fairly wide range of elderly people living in the community. Clark and Anderson culminate their work with the delineation of five specific adaptive tasks required for successful aging. The five tasks are 1) recognition of aging and definition of instrumental limitations, 2) redefinition of physical and social life space, 3) substitution of alternate sources of need-satisfaction, 4) reassessment of criteria for evaluation of the self, and 5) reintegration of values and life goals.

The first adaptive task above refers to the individual's acceptance that there are normal age-related changes that will occur and to which one can successfully adapt. These changes include psychosocial as well as physical change. For example, a person may experience retirement, loss of friends by death, or loss of family members. A person who is unable to cope with inevitable changes in aging are considered to fail in the completion of adaptive tasks. Second, the redefinition of the physical and social life space refers to the changing of certain social roles and social relationships. This is different from abandoning such roles and is also not the same as disengagement. An individual may need to change the criteria for success so that they will not continually frustrate themselves by having expectations or demands placed on them that they are unable to meet. The third task of substituting alternative sources for need satisfaction refers to a willingness to change to equally satisfying activities while giving up those that are now overly demanding (e.g., club involvement, physical health regimens, hobbies, or other self-placed demands which have become depleting rather than augmenting). The fourth task refers to the need to change the criteria for the evaluation of one's self concept. In some activities, such as sports, accommodation to changing athletic abilities is achieved by organizing competitions by age grades. Likewise, the reference point by which the individual will judge himself as an old person must be altered. The fifth task refers to the organization of the previous suggested changes into a system which is at once functional and personally satisfying. The individual who is successfully aging will not only be individually satisfied with their life position, but will articulate their life with other social spheres.

In summary, Clark and Anderson suggest that successful aging requires a shift to secondary values of American culture which have been present during their life but which have been selectively ignored until it becomes essential to use them. The adoption of the secondary values leaves the primary values aside without sacrificing one's personal sense of worth. Clark and Anderson (1967:429) describe the values shift as "...conservation instead of acquisition and exploitation; self-acceptance instead of continuous struggles for self-advancement; being rather than doing; congeniality, cooperation, love, and concern for others instead of control for others."

The elderly occupants of single room occupancy (SRO) hotels appear at first glance to be disenfranchised loners who are perhaps pathologically unable to maintain normal social relationships. However, research by Jay Sokolovsky and Carl Cohen (1983) shows that these elderly individuals do have functional personalities and social networks which operate in systematic and unexpected ways. Using anthropological fieldwork methods and social network analysis, Sokolovsky and Cohen (1983) showed that 1) most of the SRO elderly occupants have sophisticated personal networks and many participate in complex interpersonal relations which maximize their successful adaptation to the urban environment, and 2) these elders are not cut off from their former associates. Instead, elders have networks which include people they have known for up to twenty years, but with whom they have interacted sporadically within an annual cycle. Older women in this sample were as successful as male counterparts in developing and maintaining social support networks.

Sokolovsky and Cohen (1983) found three network characteristics which represent functional adaptations to maintaining health and well-being in late life in the urban environment. First, Sokolovsky and Cohen found that social networks of the SRO occupants were not evenly distributed between kin and non-kin systems with which they regularly interact. The social network is typified as a "cluster network" in which small clusters of network members exist distinctively from other clusters. From a given vantage point, a person makes differential use of these clusters based on the context of need or other purposes for interaction. One significant function of this "diffuse-cluster configuration" is the avoidance of creating reciprocal obligations in the larger social network. Next, Sokolovsky and Cohen note that such highly personal and visible actions, like giving direct aid to a cohort member to promote health and well-being, is done with a minimum of visibility and sometimes with explicit denials that the subject has any intimate social ties or dependency on anyone. The intimate ties that do exist are the result of long term friendships that have persisted from job or family settings into the present. However, despite the claim that no one "interferes" with anyone else, Sokolovsky and Cohen have noted numerous acts of caregiving among SRO occupants. Third, Sokolovsky and Cohen note that a simple head count of active social support network members would include very few people. However, the system of selective use of the clusters provides the key to understanding how the social relationships function for the SRO resident and how support group members with whom no contact has been made for many years are still considered critical support agents. Adaptation is again the key perspective in understanding the life ways of the SRO elderly.

Like the above study, another one that reveals the counter-intuitive was done by Robert Rubinstein (1985) in a literature review of materials concerning elderly people living alone. Rubinstein's analysis uses a three-part framework consisting of "antecedents of living alone", "social correlates of living alone", and outcome variables such as "living alone and its consequences for well being." First, the elderly person's chances of living alone in late life is increased among women especially if they have had few children. Also, differential survival of females to males means that more women will live alone. Also, the never married and those divorced late in life will experience life alone at greater frequencies than others. Because American values of independence and fear of dependence are quite powerful, many old people resist efforts to bring themselves to the households of family or friends because they "do not want to be a burden."

The social correlates to living alone can be approached from a simplistic viewpoint of simple head counts of support network members. However, the more revealing approach is one similar to the Sokolovsky and Cohen (1983) approach in which function and meaning is emphasized. This allows for the revelation of positive as well as negative aspects of support network membership. Rubinstein also points out that social support can be unrelated to specific events and more related to linkages based on family, friendship, and proximity of network members. "Event-related" social support includes events such as bereavement and illness as antecedent factors of interaction among support group members. Rubinstein makes the distinction of "support in-home/out-of-home" as critical for elderly people living alone compared to those who have co-residents. Rubinstein reports that there is conflicting evidence regarding the effects of living alone during bereavement except to point out that only those who live with someone else have the effect of immediacy of a co-resident for support during bereavement. Likewise with health crisis, the proximity of a caregiver assistant fosters the immediate meeting of needs. Even someone who becomes ill and has benefit of a large support network outside the home may experience gaps in needs and fulfillments.

The outcome of living alone relative to well being is examined along four dimensions (Rubinstein 1985). The first dimension refers to the effects of health on the well being of those living alone. Rubinstein (1985) reports that a person's health and related functional capacity are strong determinants of a change in living arrangements. Generally, the poorer the health the more likely the risk of moving to another arrangement, typically that of an adult child or of an institution. Next, the effect of childlessness is felt in late life by higher levels of dissatisfaction with family life among those couples who have no children. Rubinstein summarizes the literature on the effects of gender relative to well being of those living alone by showing that widowers are more likely to experience stress as well as mental health problems, social isolation, mortality, and suicide. This phenomenon is attributed to sex role behavior over the lifespan which may poorly prepare the male for the nurturance and maintenance of social relationships.

Fourth, the direct effects of marital status on the well-being of older persons living alone produces a mix of findings. For some, there is reduced stress and development of reciprocal obligations resulting in an increased satisfaction with life. On the other hand, living alone and experiencing bereavement puts one at high risk of depression. Late life divorce is also related to disenchantment with one's life circumstance. Lastly, Rubinstein reports on the well being of those living alone as mediated by social relations. Because loneliness and social isolation are not necessarily co-dependent, the older person living alone may only appear to the outside observer as lonely, while in fact having a functioning network at their disposal. This approach is reminiscent of the Sokolovsky and Cohen (1983) field research of SRO occupants. However, the possibility of low life satisfaction may occur in the absence of a co-resident or ineffective social network. It is clear that the use of "loneliness" cannot be automatically applied to a person who is living alone in late life.

*Age-Segregated Non-Institutional Community Studies*

Most older adults residing in separate residential units in a community are reasonable healthy. But, what is "health" for these ambulatory community dwelling elderly? Health is not the absence of the disease. More often it is necessary to dissect "health" from a monolithic concept into a multifaceted, dynamic phenomenon as exemplified by the several models of health delineated earlier.

In late life, mental health is maintained in part by maximizing functional capabilities regardless of the number of entries on a diagnostic roster. In the parlance of occupational therapy, a person who maintains the ability to successfully perform the usual activities of daily living (ADLs) is highly functional. In terms of health in general, a broader adaptation beyond ADLs is required for developing a sense of independence, self-determination, and high level health perception. That is, coping is more a mental state (i.e., a culturally constructed perspective or belief operating in the context of relevant positive or negative values) than a physical achievement.

In the research examples below, the themes of adaptation, mental health, physical health, economic, and social network affinities all interact to produce a subjective perception of positive or negative health status.

One of the first studies of villages built specifically for older people was done by Jerry Jacobs (1974) in *Fun City: An Ethnographic Study of a Retirement Community*. Jacobs used an ethnographic fieldwork methodology to describe the way of life in Fun City, a community of 5,700 people over the age of 50. Fun City is located in the southwestern United States and has an average resident age of 63 years.

Jacobs reports that only a small number of people participate in active ways of life including shopping, participating at the activity center, involvement in hobbies and sports, and involvement in service, fraternal, and political clubs. These residents are the "visible minority." On the other hand, the "invisible majority" of residents at Fun City lead a more hidden passive way of life. Most of these residents pass their time with activities that take place within a household setting. Health problems and functional impairments account for the low level of external community activity by the more passive Fun City residents. Another reason for non-participation is that professionals who have retired to Fun City are "besieged" by others in social settings for their professional opinion without billing. Therefore some of the professionals choose not to place themselves in such circumstances.

Adaptation to the village of elders by Fun City residents is often one of limited social interaction to preserve a sense of Fun City as a "...nice, friendly, safe place to live" (Jacobs 1974:36). It is important that residents not give offense. This community-based perspective is similar to Goffman's institutional-based perspective of adaptation in which a person follows all of the expected demands placed on them thereby producing a sense of obedience and congeniality to institutional dictates. This is Goffman's "colonizer" and Jacobs "low-key" resident who uses non-aggressiveness as an adaptive strategy.

Jacobs considers Fun City to be successful in terms of its marketing strategies, but Fun City has proved to be a "false Paradise" because age-segregated retirement settings are radically different than the mixed natural settings in which people have been living for the majority of their life. According to Jacobs (1974: 83), Fun City represented a challenge to the older person's psychosocial adaptive resources because they were searching for and expecting a paradise. But, when the many ideal services and activities were provided, it was experienced as an "alien thing" which produced discomfort, that is, "dis-ease" (Antonovsky 1979).

It is well known that the peak earning age period for Americans is in the middle 50's after which spendable earnings decline. The usual high cost of retirement to Florida for example, involves the purchase of some land and a dwelling unit. For many retirees, the answer is found in manufactured housing. Michael Angrosino (1976) conducted a study of a mobile home community in an urban city on Florida's west coast to determine how the residents adapted to their community. The mobile home community is known as Stoney Brook Park and is owned by the city which has imposed admission criteria based on minimum age of 62 and/or some disability. Stoney Brook Park is the home of about 400 elderly residents.

Stoney Brook Park is composed of people who have spent an average of 21 years in Florida rather than a group of recent migrants. Also, between 50-75% of the people own and drive their own cars. The major anxiety of the residents of Stoney Brook Park is insufficient funds to pay for necessary medical services. The interaction of high health services need, fiscal limitations, and produced anxiety is stark. The residents of the Park have developed several activities for their own interests during the first five years of the Park's existence. These include a Golden Age Club which sponsors social events, a hot lunch program, dances, and bingo. Other activities not sponsored by the Park but of importance to the residents include shopping and church attendance. However, most people rated watching TV as a frequent and important past time.

Angrosino sees the Park as an "administered community" defined as a social setting in which others have a large decision making capacity over the residents. However, it is also shown that the concept of a social community is not what has developed at Stoney Brook Park. Stoney Brook Park is "...an artifact of political and social authority exercised by outsiders" (Angrosino 1976:179). The result is a sense of frustration, low self-esteem, and apathy among the residents. These are not the prerequisites for high level health. Angrosino notes that in this age-segregated community the users have little more investment in their residence than native Americans housed on reservations feel any "warmth of attachment" to the governmental bureaucracy determining their destiny.

In a 200 unit high-rise funded by the city of Milwaukee, Eunice Boyer (1980) examined the health perceptions of older people as a function of their level of activity within the housing project. Boyer confirmed the relationship of high quality health perception related to a high level of activity and social interaction. She also demonstrated that the elderly residents of this public high-rise are quite vital and functional in their creation of social networks and in the development and response to formal leadership within it. Boyer also reports that the homogeneous composition of residents (i.e., relatively well elderly) contributes to the further sense of health and well-being of residents by preventing clashes with younger and perhaps better educated residents who might more easily rise to informal power.

In the same Milwaukee public housing project that Boyer investigated, Jonas and Wellin (1980) examined the development of residents' mutual support networks which can buffer or prevent morbid episodes (McKinlay 1980). The structure of these support networks are influenced by gender and familiarity among the network members. They report that women are likely to engage in the "mother-hen pattern" of giving aide which is characterized by intimacy and long term relationships, as part of a pattern of generalized reciprocity. In contrast, men practice more a type of balanced reciprocity described as a "customer pattern" where reciprocity is immediate and personal relationships or intimacy is less important. Negative reciprocity emerges only if a person continues to place demands in the absence of a clear place in the social network or intent of reciprocal action. Jonas and Wellin (1980) dismiss the validity of the dumping ground or warehouse concept for the elderly people of the public housing project examined by them. They perceived more a total configuration of positive and healthy adaptation to their particular late life context.

Haim Hazan (1980, 1982) conducted research in the East End of London in a predominantly Jewish day care center. A day care center was established in their neighborhood with a unique non-interference policy in the lives of the users or in the organization of the center's activities. These elderly people seized upon an opportunity to create their own organization and social system rather than maintain an amorphous daily existence at the center. This strategy recalls the Antonovsky (1979) "sense of coherence" as a prerequisite for health. The day care participant was able to arrive and leave at will, pursue interests and activities at will, and develop new activities as deemed necessary.

A primary ingredient leading to the high level of activity and satisfaction with this day care center is found in the cultural conception of help and caring. This population of elderly Jewish immigrants attached enormous importance to helping others even to the exclusion of expected reciprocity.

While this population experienced severe social and community disruption at the end of World War II due to the bombing of London, Hazan shows that the disruption and near disintegration were functional prerequisites for redevelopment under the impetus of available welfare services. These elderly immigrants seized upon their common ethnic identity and life events, transforming them into a well-spring of reintegration and readjustment (cf. Simic 1985). Hazan (1982:358) states, "... under certain circumstances, what might be viewed as a step toward termination and gradual conclusion of the life cycle is merely a pre-condition for social revitalization and self-realization."

## Studies of Institutional Age-Segregated Communities

The first image that comes to mind for many people when referring to age-segregated communities is the nursing home. In fact, anthropological interest in nursing homes is one of the earliest endeavors of contemporary anthropological research with older people. Institutionally confined individuals do represent a captive and often subjugated research population. However, in recent years ethical standards of access to research subjects and research sites makes the nursing home patient more autonomous in their decision of whether to participate as a subject in an ethnographic investigation.

There is also a social imperative to thoroughly study nursing homes because almost 39% of the elderly population in this country will live some portion of their late life in a nursing home facility (Vicente, Wiley, and Carrington 1979). As more and more people survive into advanced old age (85 and older), the risk of institutionalization is increased. Also, the increased risk of dementing disease like Alzheimer's disease is associated with advanced age. Therefore, survivorship into the 80th decade carries with it the risk of institutionalization and/or the additional risk of developing dementia which may require institutional care. From a financial perspective, understanding the human organization of nursing homes and the cost of such care is mandatory for controlling the federal budget. A number of anthropologists have been involved in nursing home studies designed to reveal the process by which institutional long term care is delivered.

Jules Henry in *Culture Against Man* (1963) discussed three nursing homes, comparing them along a dimension of humane to in-humane. From his studies, Henry developed the concept of "human obsolescence." Participation-observation in the three homes revealed that regardless of the intent to provide quality care or even to ignore quality care, the noxious effect of institutional characteristics pervaded the life of the geriatric patient. Henry's nursing home research must be placed in the context of Erving Goffman's (1961) research in institutional facilities such as prisons and mental hospitals. The astute observations of Goffman and Henry are both oriented toward the control of the "inmate" (i.e., patient or resident) by institutional structures and phenomena. From this perspective, human adaptation is de-emphasized and the role of the institution in suppressing human vitality and creativity is highlighted. The worst of the three nursing homes investigated by Henry is a private, for-profit home, as are 79% of American nursing homes (Butler 1975). Conditions were wretched even though upper level staff commented that they loved their work. Yet, in 33 days of observation, only two baths were observed. Also, mealtimes were few and far between and marked by unsanitary conditions and bad food. Henry relies on the metaphor of a begging dog to convey the behaviors of patients and food in this nursing home. Henry notes that acquiescence and reminiscence are adaptive traits to the noxious environment of this nursing home. The residents are characterized by Henry as powerless because they have no money or friends, and, by social definitions they are a mixture of animal (dog-like), the child, and the lunatic. Psychically, reminiscence about the past allows the patient to regain a sense of purpose and life-meaning. However, fear pervades the present because those who complain too much are at risk of restraint or expulsion.

In the nursing home reported as the least noxious of the three, Henry (1963) discusses the "national character" of the facility in an effort to convey the institutional life experience under what appeared to be a state-of-the-art home. Henry reports that the staff is kind in its intent, but maintains an attitude of indulgent superiority over the patients who they define as disoriented children. The psychosocial needs of the patient are ignored so that staff attends to the bodily needs only. Henry summarizes this "best example" nursing home as one oriented mainly toward "... medical care, feeding, and asepsis." (1963:474). The implication is that the medical model informs the type of care given, while the psychosocial problems model, which is humanizing, is ignored (Bowker 1982).

Murray Manor, a 360 bed nursing home, that is a sectarian non-profit facility is the focus of a field study by sociologist Jaber Gubrium (1975). Therefore, due to the newness of the facility, only about one-third of the beds were filled. Within two years, the census was at about 60% of capacity. From the sociological perspective, Gubrium examines the nursing home and its organizational structure by delineating the social/behavioral constituents of the nursing home care process and how they interact. This includes the administrative staff, interrelationships between the residents and patients, the work of the nurse aides, strategies for passing time by the residents and patients, and the way in which death was handled by patients and staff.

Gubrium finds that the top staff (the administration) are well insulated from the patient population and is unaware of the distance between themselves and the life experience of their patient population. The administrative staff considers the nursing home to be operating well when there are few complaints from patients, nurse aides, and families. When things go wrong, it is not the administrative structure that is to blame, rather it is people who are being "too individualistic" and are not conforming to the goals of the organization. He finds that decisions made by administrators do not benefit from an understanding of the social interaction complexities of nursing home life from the patient perspective. Patients are typically not involved in making administrative decisions, having an impact on administration, or even having access to the space or place occupied by administrative staff. These factors maintain the insulation between those who manage the lives of the patients and those who are managed.

The bulk of any nursing home employee group is the nurse aide staff. In fact, 43% of the employees in nursing homes are nurses aides (Subcommittee on Long Term Care 1975). The nurse aide constitutes Gubrium's "floor staff." The floor staff exercises a balance of control between the top administrative staff and the patient population. The nurse aide staff is invested in maintaining a work style designed to minimize interference from both administration and patient. Thus, the nurse aide staff wields considerable informal power in the daily delivery of long term caregiving.

Passing time is a task for the nursing home patient. The patient is not passing time for the achievement of some goal such as discharge nor are they passing time simply to get from one point on the clock to another. The tracking of time occurs in two ways. One obvious method of tracking time is to use the clock. The other time-tracking method is observation of the nursing home staff work activity as it occurs in its daily routine. But, Gubrium also examines ordinary time passing events such as eating, sleeping, walking, watching, talking. Rather than simply noting the presence of these ordinary activities, Gubrium examines the meaning that each one of these ordinary events has for the patient population. For example, very detailed discussions about food service is fare for daily conversation. Conversations are organized around pre-meal topics, arrival of food on tables, the quality of the food itself, and a review of past meals and future possible menus. Furthermore, each floor has its own characteristic pattern of dining room use. On the first floor for ambulatory, high functioning patients, food consumption occurs in about 15 minutes, but the discussion about the meal itself is much longer. Even sleeping is analyzed in terms of the difference in place and the amount of time spent sleeping. Gubrium notes that a person who spends most of the day sleeping in a bed is negatively evaluated by others but the same person sleeping in one of the nursing home lounges is more positively evaluated by others because it is considered just dozing off.

The nursing home as presented by Gubrium is a special community with all the complexities of daily interaction and life just as much as other life spaces experienced in their younger years. The perspective that nursing homes are filled with stagnant and frail elderly people is not an accurate reflection of nursing home life. Gubrium also provides a different perspective from Henry (1963) who relied on the impact of the "total institution" as a determinate of misery of the patients. Gubrium shows the impact of the institutional aspects of Murray Manor but also reveals the active dynamics of the patient and staff population all of whom make creative adjustments to their circumstances. Goffman (1961), with whom the concept of institutional totality is often associated, in fact addresses in detail the niches for creativity and resistance that are found in the institutional regimen and which are routinely exploited by workers and patients.

At Pecan Grove Manor Nursing Home, J. Neil Henderson (1981) examined the interactions between nurse aides and patients in the discharge of long term caregiving and receiving. Henderson used an ethnomedical perspective to analyze performance characteristics of the nursing home staff. The purpose was to reveal the actual job performance characteristics regardless of the officially assigned job title and work description. Henderson credits the ethnomedical perspective for revealing a hidden therapeutic function of a worker outside the designated boundaries of the health care staff: the housekeeper.

The housekeeper was revealed as an indigenous psycho-therapist who provided specific psychosocial support to the patients. Although the nursing staff is officially charged with responsibility for health care matters, the licensed nursing staff and paraprofessional nurse aides discharge "basic care" needs which means they are responsible for "bed and body care" (cf. Gubrium 1975; Friedson 1970). The psychosocial dimension of life is not considered a primary nursing task.

Housekeepers were able to function as indigenous therapists partly because their workstyle kept them in the patient's room for a sustained period of time. In order to do an effective job of cleaning, the housekeeper will be in the patient's room for almost 20 minutes. By contrast, the nurse aide is rapidly discharging the bed and body duties from patient to patient and responding to call lights on an ad-hoc basis. Total contact time with patients may be longer for aides than housekeepers over an 8-hour shift, but aide work is episodic and not sustained interactional time.

Also, the housekeeper wears ordinary street clothes and are not identified as caregivers to patients. The patient-caregiver hierarchical distinction is in this way lost and is replaced by adult to adult interactions. Although the psychosocial support function of the housekeeper is extremely important to the well-being of the patient, the other staff groups are unaware of the multidimensional role of the housekeepers. An episode of significant psychosocial support is an invisible cargo to casual observers. Such "therapy" is not showy, does not involve mechanical instrumentation, but demands a personal demeanor which is conducive to empathic conversations and the sustained time for such discourse to develop. Henderson summarizes his findings by saying "... while nurses tend to the body, housekeepers tend to the mind" (Henderson 1981:303).

Cross-cultural research in nursing home settings remains uncommon. However, Jeannie Kayser-Jones (1981) has conducted comparative research in Scottish and American nursing homes. While the two are not true equivalents because of different overall health care systems, the primary task of taking care of frail older people has common dimensions which can be compared across systems and cultures. Kayser-Jones uses an exchange theory approach to examine the adaptive capacity of patients in both the Scottish and American nursing home. Exchange theory as applied to the institutionalized aged involves an imbalance in social power because many patients are dependent upon the staff for resources to create some semblance of control, power, and honored identity. However, compliance with staff demands is often obtained because the patients have few resources with which to reciprocate. Thus, an imbalance of social power is perpetuated to the detriment of the patient population.

Kayser-Jones considers the Scottish patient to have a superior nursing home experience to the American patient because they have more resources with which to reciprocate with the staff. Because of the national health care system of Scotland, the elderly institutionalized patient has greater access to quality medical and nursing care, quality activities, and a dedicated nurse aide staff. This is less common in the United States. Nonetheless, in the American nursing home, some older people manage through their own creative efforts to develop a balanced social relationship with some of the nursing home aides. For example, Kayser-Jones reports the case of an American patient who is able to mend and alter clothes for one of the nurse aides who reciprocates by doing some small shopping for her. The patient comments that she would never charge for her work. However, the nurse aide reciprocates by helping the patient with an activity that is outside the actual job description for the nurse aide.

Much of the superior care in the Scottish nursing homes is attributed to the socialized health care systems of the United Kingdom in contrast to the proprietary nursing home industry of the United States. The outcome for the institutionalized elderly in America is a nursing home experience with low quality medical care, minimal services so that overhead can be cut to increase profits for the business, poorly paid nursing staff, and adherence to the medical model for the long term management of incurable chronic disease. Thus while patients adapt to nursing home environments in remarkable ways, the adaptation of the nursing home industry to the business of long term care and government regulation serves to maintain a social and economic imbalance in favor of the nursing home industry.

The core concept of anthropology is understanding the influence of culture on human existence. Maria Vesperi (1983) examines the low quality of life in American nursing homes from the perspective of the American cultural concept of old age and its relationship to the wish for a cure of health care problems. Vesperi argues that administrative policy in nursing homes is directly informed by the concept of old age and what old age means, so that old age is perceived as a time of dependence with a related need for paternalistic management of one's life. Also, the nursing home patients are perceived as inadequate to evaluate their own life circumstances, discuss medical intervention options, and make other decisions about their experiences in the nursing home, such as food selection or movement from one room to another. Moreover, it is assumed that withdrawal (like "disengagement" was viewed in the larger society) is a normal and desirable response to aging. Therefore, any patient response which is active and contrary to passive compliance such as wandering away, verbal retaliation, or attempts to manage one's own destiny are considered inappropriate and may be labeled signs of cognitive impairment. The latter engenders a host of possible management strategies including physical restraints, chemical restraints or psychosocial retaliation.

Nursing homes as a part of the American cultural environment have been examined by Colleen Johnson (1987; Johnson and Grant 1985). American values of self-determination and independence can be completely undermined by the problems of chronic disease which often necessitates nursing home placement. Furthermore, nursing home placement is defined by many as the warehousing or abandonment of loved ones which is in opposition to humanitarian values. Even some purportedly objective analysis by gerontologists and other social scientists are revealed by Johnson as subtly reflecting the American values of self determination, independence, and optimism by overlooking the plight of the frail elderly person and constructing an overly optimistic, misleading view of the aged by selective use of data. Also, Johnson examines the conflict of the high value placed on biomedicine which glaringly fails in the face of long term, chronic, incurable diseases which many nursing home patients experience. Johnson also sees the origin and development of the nursing home as consistent with the Western societal proclivity to found institutions for those with stigmatized afflictions "... not to suppress them, but to keep it (the affliction) at a safe distance" (Johnson 1987). In spite of perpetuating institutional solutions, Johnson considers the nursing home an important resource, albeit one of last resort, because the American nuclear family system has difficulty managing adult members who have become dependent particularly within a health care system which favors acute care.

*Disease-Specific Studies*

In addition to community studies, a trend emerging in anthropology and aging identifies diseases common to the older person as the principle subject of study. Such investigations examine disease both from a cultural perspective and within the context of the late life experience. Recalling the earlier discussion of unique aspects of geriatric patients, it is reasonable to examine disease in older adults not only from the changed physiological perspective but also from the changed psychosocial perspective in the context of cultural meaning. While limited literature on disease-specific studies reflects the newness of this trend, materials on two diseases will be reviewed here. These are dementia such as Alzheimer's disease and cerebrovascular accidents or strokes, both of which occur at great frequency among older Americans and have received research attention by anthropologists. Also, it is important to note that other disease-specific research has been conducted by anthropologists, for example Linda Mitteness' research on urinary incontinence (Mitteness 1987a, 1987b) and Lois Grau and Deborah Padgett (1988) on depression.

*Dementia*
Dementia is the fourth leading cause of death nationally. Cross-sectionally about 4% of the 65 and over population will suffer from a dementing disease. However, if age cohorts are examined the 75-80 age cohort will have a prevalence rate of 10-15% and the 80+ cohort will have a prevalence rate of 20-

25% (Mortimer 1983; Mortimer, Schuman, and French 1981). Alzheimer's type dementia is the most common type of dementing disease constituting 50-60% of all cases (Reisberg 1983). Strokes or cerebrovascular accidents occurring in sufficient numbers in the brain can produce a vascular dementia accounting for 15% of all cases of dementing disease (such as multi-infarct dementia). Also, Alzheimer's type dementia and vascular dementia can be present in one individual which is the case in 25% of all dementing patients.

Enormous expenditures of physical, research, and human resources are being used in the treatment or management of dementing disease (Jau and Ernst 1987; Weiler 1987). Much of the research is conducted in the biomedical and clinical arenas. However, lagging far behind is analysis of the sociocultural correlates of dementing disease (For other literature see Gubrium 1986, 1988; Stafford 1982; Valle 1987). Henderson (1987) focuses on dementia in late life not only from the pathophysiological and clinical viewpoint but from the "illness" perspective as well. After discussing common mental health disorders among the elderly, the "dementia epidemic" is examined both by a sociomedical history of Alzheimer's disease and a discussion of diagnostic criteria for Alzheimer's type dementia.

However, the central focus is on sociocultural management issues using research in a group of Latin families as a window into sociocultural impact on the experience of dementia. Henderson (1987) discusses several factors that are embedded in the general culture of American society that worsen the impact of Alzheimer's disease on patients and families. The medical model of care is seen as a cultural product in which diagnostic taxons are constructed and reconstructed over time. "Senility" as a concept is still a central diagnostic taxon in the minds of many practitioners and is tantamount to a non-treatment posture by many. Also, the medical model is characterized by a unidisciplinary treatment approach which fails to yield optimal treatment in a bioculturally complex and chronic disease process. Thus, chronic disease is experienced by the family members and patients in a medical culture oriented to in-patient care for short periods of time. These facilities are operated by practitioners with little geriatric or interdisciplinary team management skills.

The American values of self-determination, independence, and free will are all undermined by failure of the brain as an organ of adaptation, interpretation, and action. In concert with this are a set of sociocultural patterns related to diffuse kinship structures, complex migration patterns, and values regarding the division of labor and sex role performance in terms of caregiving dynamics. Values of independence and neolocal post-marital residence patterns often cause the burden to be placed on one central caregiver although many secondary caregivers may be available.

Traditional divisions of labor based on sex can be seen in terms of a cohort effect in which the older American of today experienced sex role socialization during a time in which discreet sex role patterns were common. In late life, if a wife experiences Alzheimer's disease, the husband usually will become a primary caregiver who may possess a very meager role repertoire of domestic duties. Such men may experience a variety of psychological conflicts resulting in sub-optimal care of the demented spouse and their own dysfunction. Future cohorts of elderly who experienced a more diffuse sex role socialization may bring a more balanced caregiving capacity to the in-home management of the dementia patient.

The failure of the medical system to address chronic aspects of disease has resulted in the spontaneous emergence of dementia-specific mutual aide societies. While many "support groups" exist, one of the fastest growing is that for caregivers of Alzheimer's disease patients. These support groups or "disease clubs" function socially as fictive kinship groups. Support groups have developed in a culture characterized by fragmented kinship networks, a failed acute care medical system and in a vacuum of knowledge regarding in-home long term care of chronic disease.

Lastly, Henderson (1987) also observed that Alzheimer's disease support groups are largely attended by Anglo middle-class caregivers. Even in communities with significant ethnic populations, the striking ethnic and class homogeneity of support group users was the norm. Field research was initiated in the Latin community of a city on the west coast of Florida to determine if support groups were culturally appropriate in this minority community. If so, how might such an intervention be effectively designed and implemented in these communities? Henderson (1987, 1989) discusses examples from a Latin Alzheimer's support group stemming from the field research. Briefly, several distinctive features of the Latin caregiving style were revealed in support group meetings and during in-home interviews. For example, the Latin caregiver differentially uses available community services by age cohorts. For example, the Latin older generation is less likely to use community services compared to the younger Latin generation. However, the older generation will make use of services when brokered by the adult children of the caregiver. Also, there is a very definite sex bias pattern for responsibility of care. While women are the expected caregivers in many societies, in this Latin community the expectation is extremely intense. In fact, Henderson and Garcia (1986) show that selection preference for female caregivers will change from kin to non-kin before using males within the family group.

Since dementia produces behavioral aberrations, the stigma of having a family member who is "crazy" is commonly discussed by the Latin caregivers. Caregivers develop a standard proactive stance by telling others that this disease is an organic disease and not a psychiatric or psychological one. Also, the support group in the Latin community is conducted in Spanish. Prior to the project there were almost no Latin members of the community attending the existing community-wide Alzheimer's disease support group, whereas now some 65 cases have been found and a regular, ethnic-specific support group meeting is well attended.

## Cerebrovascular Accident

Sharon Kaufman (1988) has followed 102 stroke patients for one year post-stroke. The focus is on the illness aspect of strokes which Kaufman identifies as the patient's subjective response to the symptoms. Kaufman demonstrates that the rehabilitation practitioners and their patients interpret the entire health care experience very differently.

Through participant observation and intensive interviewing, Kaufman demonstrates that the practitioner of rehabilitation medicine and other allied health disciplines direct their efforts toward the rehabilitation process as revealed by their concern for observation, measurement, and constructing definitions for success of therapy. However, patients with stroke-induced brain damage perceive their task in rehabilitation as one of recovery. Recovery is a diffuse goal related to subjective beliefs of normality, continuity with one's previous life experience and capability, and consistency with one's self identity.

Kaufman shows that even if the practitioner criteria for success are assimilated by patients in the early stages of rehabilitation, divergence may appear later in treatment as patients recover more function and adapt better to their condition. The result is a re-evaluation of the criteria for progress and of their perception of self in the future. Kaufman (1988:85) discovered three areas into which patient concern falls: 1) discontinuity of life patterns, 2) the failure to return to normal, and 3) the redefined self. These issues are not held by the practitioner group. For example, even those patients who have "perfect" performance in therapeutic work will inevitably report that they still have not returned to their pre-stroke sense of normality.

Kaufman presents two detailed cases. In both, the patients are elderly women who have suffered strokes with permanent paralysis of the left side. Yet, one patient assimilates the rehabilitation model and adapts her own individual goals and identity into the process, while the other patient does not accept the rehabilitation process and is, in medical terms, non-compliant. Kaufman (1988:97) interprets these two cases as an example of "... the subjective experience of recovery, which may or may not resemble the structured goals of stroke rehabilitation." For the compliant patient, recovery was considered within the framework of an ability to regain function in the activities of daily living and to achieve a sense of continuity of self from pre-stroke to post-stroke days. On the other hand, recovery for the other patient meant an ability to avoid the therapeutic interventions designed to bring about an improvement in her condition. The latter patient strived to maintain the same independent living after the stroke that she had achieved before it.

Kaufman (1988) outlines the implications of her research in four areas. First, she notes that practitioners must be aware of the biographical (life) history of their patient. As Kaufman (1988:97) states, "In order fully to understand the meaning of stroke to the individual who experiences it, one needs to know who that person is." While a careful taking of a patient's history is standard medical practice, the history of the patient's life must be particularly detailed and values about health matters and life and death must be elicited in the arena of chronic disease and long term care. Second, the issue of compliance must be understood in terms of the stroke patient's beliefs about the rationality of treatment and the life circumstances of the patient at the time of the stroke. Practitioners who measure compliance based exclusively on the patient's adherence to their treatment protocols may negatively judge the patient when they develop their own, creative ways of adapting and making use of the therapeutic recommendations.

Third, Kaufman examines stroke from the perspective that two frameworks in medical care exist and represent a departure from the classical biomedical tradition. The "biopsychosocial" approach attempts to create a multidisciplinary perspective regarding disease and its treatment. And there is "Medicalization" in which psychosocial matters are converted to medical matters and thus are subject to medical intervention and interpretation. While Kaufman (1988) sees these two processes as having some merit, at the same time they carry the risk of turning non-medical arenas of life over to control by physicians.

Lastly, the elderly population is more vulnerable to health care mismanagement than younger and middle aged populations, and for many reasons. There is a high expectation for health care intervention stemming from the belief that the management of the older person is medically required, based on the belief that older people have more chronic conditions and more hospital days than do younger age groups. However, presence in a hospital does not imply that psychosocial dimensions of life have been abandoned or are unimportant to the geriatric patient. Overall, Kaufman's research reveals that the older stroke patient is still a creative person and can have an active part in the health care system. Practitioners and their clients can achieve rehabilitation or recovery if the subjective personal meanings of their afflictions are examined and integrated into treatment regimens.

CONCLUSION

The contribution of anthropology to health and aging is extensive. Anthropologists have investigated the late life experience in diverse settings ranging from the elderly living in communities to hospitals and nursing homes. Yet, more work needs to be done.
Research in health and aging by anthropologists is derived largely from medical anthropologists with research interests in aging. As the article presented here shows, much of this work has been oriented toward the sociocultural organization of old age relative to health, disease, and well being. This information has provided critical data for understanding the ways in which old people respond to changing life and health conditions. However, the biocultural framework of anthropology in general and specifically medical

anthropology can bring a powerful analytical and applied orientation to the study of aging.

The biocultural perspective can integrate the normal and pathological physical changes of late life into the socioculturally influenced illness response. A full analysis of any given late life experience must include not only the sociocultural dimension but the changing bodily vehicle which mediates the actual behavioral response.

Also, the culture of geriatric medicine and other geriatric specialities has not been sufficiently investigated by anthropologists with gerontological interests. To some degree, sociologists have preceded anthropologists in the examination of the clinical encounter from a cultural perspective. However the official corpus of training literature and clinical practice concepts in geriatric medicine remains to be elucidated by anthropologists.

For decades, medical anthropologists have investigated specific human responses to physiologic disease. However, the focus of cultural response to physical disease as it pertains to the elderly population has received little attention. Many policies related to national health programs that affect the elderly are related to presumed health and functional changes experienced by the elderly in late life. The bioculturally-oriented anthropologist studying age would produce an extremely important body of information integrating geriatric biology with sociocultural experience and vice versa. These avenues in aging research will grow in the next several decades as the number of elderly survivors continues to rise. Anthropologists have made a good start into the gerontological field and now need to target their research activities to a more bioculturally oriented framework.

## REFERENCES

Adams, F. M. (1972) The Role of Old People in Santo Tomas Mazaltepec. In Aging and Modernization. D.O. Cowgill and L.D. Holmes, eds. New York: Appleton-Century-Crofts.

Alland, Jr., A. (1966) Medical Anthropology and the Study of Biological and Cultural Adaptation. American Anthropologist 68:40-51.

Angrosino, M.V. (1976) Anthropology and the Aged: A Preliminary Community Study. The Gerontologist 16:174-80.

Antonovsky, A. (1979) Health, Stress, and Coping. San Francisco: Jossey-Bass.

Arth, M.J. (1968) An Interdisciplinary View of the Aged in Ibo Culture. Journal of Geriatric Psychiatry 2:33-39.

Berkman, L.F. (1981) Physical Health and the Social Environment: A Social Epidemiological Perspective. In The Relevance of Social Science for Medicine. L. Eisenberg and A. Kleinman, eds. Boston: Reidel.

Besdine, R.W. (1982) The Data Base of Geriatric Medicine. In Health and Disease in Old Age. J.W. Rowe and R.W. Besdine, eds. Boston: Little, Brown and Company.

Bowker, L. (1982) Humanizing Institutions for the Aged. Lexington, Massachusetts: Lexington Books.

Boyer, E. (1980) Health Perceptions in the Elderly: Its Cultural and Social Aspects. In Aging in Culture and Society. C. Fry, ed. Brooklyn: Bergin.

Brenner, M.H., A. Mooney, and T.J. Nagy (1980) Assessing the Contributions of the Social Sciences to Health. Boulder, Colorado: Westview Press.

Brody, J.A. and D.J. Foley (1985) Epidemiologic Considerations. In The Teaching Nursing Home. E.L. Schneider, ed. New York: Raven Press.

Butler, R. (1975) Why Survive?: Being Old in America. New York: Harper & Row.

Byrne, S.W. (1974) Arden, An Adult Community. In Anthropologists in Cities. G.M. Foster and R.V. Kemper, eds. Boston: Little, Brown and Company.

Clark, M. (1973) Contributions of Cultural Anthropology to the Study of the Aged. In Cultural Illness and Health. L. Nader and T. Maretski, eds. Washington, D.C.: American Anthropological Association.

Clark, M. and B. Anderson (1967) Culture and Aging: An Anthropological Study of Older Americans. Springfield, IL.: Charles C. Thomas.

Clark, M.M. and M. Mendelson (1969) Mexican-American Aged in San Francisco: A Case Description. The Gerontologist 9:90-95.

Cowgill, D.O. and L. D. Holmes (1972) Aging and Modernization. New York: Appleton-Century-Crofts.

Earley, L.W. and O. von Mering (1969) Growing Old the Outpatient Way. American Journal of Psychiatry 125:135-139.

Eisenberg, L. and A. Kleinman, eds. (1980) The Relevance of Social Science for Medicine. Boston: Reidel.

Engel, G.L. (1980) The Clinical Application of the Biopsychosocial Model. The American Journal of Psychiatry 137:535-544.

Engel, G.L. (1977) The Need for a New Medical Model: A Challenge for Biomedicine. Science 196:129-136.

Fabrega, Jr., H. (1975) The Need for an Ethnomedical Science. Science 189:969-975.

Fabrega, Jr., H. (1974) Disease and Social Behavior: An Interdisciplinary Perspective. Cambridge, Massachusetts: MIT Press.

Foster, G. and B. Anderson (1978) Medical Anthropology. New York: John Wiley and Sons.

Foster, G. (1975) Medical Anthropology: Some Contrast with Medical Sociology. Social Science and Medicine 9:427-432.

Friedson, E. (1970) Profession of Medicine: A Study of the Sociology of Applied Knowledge. New York: Dodd, Mead, and Company.

Fries, J.F. (1980) Aging, Natural Death, and the Compression of Morbidity. New England Journal of Medicine 303:130-135.
Fries, J.F. (1984) The Compression of Morbidity: Miscellaneous Comments about a Theme. The Gerontologist 24:354-59.
Goffman, I. (1961) Asylums. Garden City, N.Y.: Anchor Books.
Grau, L. and D. Padgett (1988) Somatic Depression Among the Elderly: A Sociocultural Perspective. International Journal of Geriatric Psychiatry 3:201-207.
Gubrium, J.F. (1988) Family Responsibility and Caregiving in the Qualitative Analysis of the Alzheimer's Disease Experience. Journal of Marriage and Family 50:197-207.
Gubrium, J. (1987) Structuring and Destructuring the Course of Illness: The Alzheimer's Disease Experience. Sociology of Health and Illness 9:1-24.
Gubrium, J. (1986) Oldtimers and Alzheimer's: The Descriptive Organization of Senility. Contemporary Ethnographic Studies. Greenwich, Connecticut: JAI Press.
Gubrium, J.F. (1975) Living and Dying at Murray Manor. New York: St. Martin's Press.
Ham, R. ed. (1983) Primary Care Geriatrics. Boston: John Wright.
Haug, M. (1981) Elderly Patients and Their Doctors. New York: Springer.
Hay, J.W. and R.L. Ernst (1987) The Economic Cost of Alzheimer's Disease. American Journal of Public Health 77:1169-75.
Hazan, H. (1980) The Limbo People: A Study of the Constitution of the Time Universe among the Aged. London: Rowledge and Kegan Paul.
Hazan, Haim (1982) Beyond Disengagement: A Case Study of Segregation of the Aged. Human Organization 41:355-359.
Henderson, J.N. (1981) Nursing Home Housekeepers: Indigenous Agents of Psychosocial Support. Human Organization 40:300-305.
Henderson, J.N. (1987) Mental Disorders Among the Elderly: Dementia and its Sociocultural Correlates. In The Elderly as Modern Pioneers. P. Silverman, ed. Bloomington: Indiana University Press.
Henderson, J.N. (1989) Alzheimer's Disease in Cultural Context. In Cultural Context of Aging: Worldwide Perspectives. J. Sokolovsky, ed. South Hadley, Massachusetts: Bergin and Garvey.
Henderson, J.N. and J. Garcia (1986) Caregiver Selection for Dementia Patients in a Latin Population. Unpublished Paper. Presented at the Gerontological Society of American meetings. Chicago, Illinois.
Henry, J. (1963) Culture Against Man. New York: Random House.
Hill, R.F., J.D. Fortenberry, and H.F. Stein (in press) The Role of Culture in Clinical Medicine. Southern Medical Journal.
Holmes, L.D. (1972) The Role and Status of the Aged in a Changing Samoa. In Aging and Modernization. Donald O. Cowgill and Lowell D. Holmes, eds. New York: Appleton-Century-Crofts.
Hughes, C.C. (1978) Medical Care: Ethnomedicine. In Health and Human Condition. M. Logan and E.E. Hunt, Jr., eds. North Scituate, Massachusetts: Duxberry.
Jacobs, J. (1974) Fun City: An Ethnographic Study of a Retirement Community. New York: Holt, Rinehart & Winston.
Johnson, C. (1987) The Institutional Segregation of the Aged. In the Elderly as Modern Pioneers. P. Silverman, ed. Bloomington: Indiana University Press.
Johnson, C.L. and L. Grant (1985) The Nursing Home in American Society. Baltimore: Johns Hopkins University Press.
Jonas, K. and E. Wellin (1980) Dependency and Reciprocity: Home Health Aid in an Elderly Population. In Aging in Culture and Society. C. Fry, ed. Brooklyn: Bergin.
Kaufman, S. (1988) Stroke Rehabilitation and the Negotiation of Identify. In Qualitative Gerontology. S. Reinhart and G. Rowles, eds. New York: Springer.
Kayser-Jones, J. (1981) Old, Alone, and Neglected: Care of the Aged in Scotland and the United States. Berkeley: University of California Press.

Kiefer, Christie W. (1971) Notes on Anthropology and the Minority Elderly. The Gerontologist 11:94-98.

Kirkwood, T.B.L. (1985) Comparative and Evolutionary Aspects of Longevity. In Handbook of the Biology of Aging. Second edition. C.E. Finch and E.L. Schneider, eds. New York: Van Nostrand Reinhold.

Kleinman, A. (1980) Patients and Healers in the Context of Culture. Berkeley: University of California Press.

Kleinman, A., L. Eisenberg, and B. Good (1978) Culture, Illness, and Care: Clinical Lessons from Anthropologic and Cross-Cultural Research. Annals of Internal Medicine 88:251-258.

LeVine, R.A. (1965) Intergenerational Tensions and Extended Family Structures in Africa. In Social Structure and the Family: Generational Relations. E. Shanas and G. Streib, eds. Englewood Cliffs: Prentice-Hall.

McKinlay, J.B. (1980) Social Network Influences on Morbid Episodes and the Career of Help Seeking. In The Relevance of Social Science for Medicine. L. Eisenberg and A. Kleinman, eds. Boston: Reidel.

Mitteness, L.S. (1987a) The Management of Urinary Incontinence by the Community-Living Elderly. The Gerontologist 27:185-193.

Mitteness, L.S. (1987b) So What Do You Expect When You're 85? Urinary Incontinence in Late Life. In J. Roth and P. Conrad, eds. Research in the Sociology of Health Care. Greenwich, Conn.: JAI Press, Inc.

Monthly Vital Statistics Report (1983) Volume 31, Number 13, October 5, p. 15.

Moore, M.J. (1987) The Human Life Span. In The Elderly as Modern Pioneers. P. Silverman, ed. Bloomington: Indiana University Press.

Mortimer, J.A. (1983) Alzheimer's Disease and Senile Dementia: Prevalence and Incidence. In Alzheimer's Disease, B. Reisberg, ed. New York: Free Press.

Mortimer, J., L. Schuman, and R. French (1981) Epidemiology of Dementia. In The Epidemiology of Dementia. New York: Oxford University Press.

Munsell, M.R. (1972) Functions of the Aged Among Salt River Pima. In Aging and Modernization. Donald O. Cowgill and Lowell D. Holmes, eds. New York: Appleton-Century-Crofts.

National Center for Health Statistics (1985) Vital Statistics of the United States, (1982). Vol. 2, Sec. 6, Life Tables. DHHS Pub. No. (PHS) 84-1104. Washington, D.C.: U.S. Government Printing Office.

Plath, D.W. (1972) Japan: The After Years. In Aging and Modernization. Donald O. Cowgill and Lowell D. Holmes, eds. New York: Appleton-Century-Crofts.

Polgar, S. (1962) Health and Human Behavior: Areas of Interest Common to the Social and Medical Sciences. Current Anthropology 3:159-205.

Press, I. and M. McKool, Jr. (1972) Social Structure and Status of the Aged: Toward Some Valid Cross-Cultural Generalizations. Aging and Human Development 3:297-306.

Reisberg, B. (1983) Clinical Presentation, Diagnosis, and Symptomatology of Age-Associated Cognitive Decline and Alzheimer's Disease. In Alzheimer's Disease. B. Reisberg, ed. New York: Free Press.

Ross, J.K. (1972) Successful Aging in a French Retirement Residence. In Successful Aging. E. Pfeiffer, ed. Durham: Duke University Press.

Rowe, J.W. and R. W. Besdine (1982) Drug Therapy. In Health and Disease in Old Age. J.W. Rowe and R.W. Besdine, eds. Boston: Little, Brown and Company.

Rubinstein, R. (1985) The Elderly Who Live Alone and Their Social Supports. In Annual Review of Gerontology and Geriatrics, C. Eisdorfer, ed. New York: Springer.

Sankar, A. (1984) "It's Just Old Age": Old age as a Diagnosis in American and Chinese Medicine. In Age and Anthropological Theory, edited by D. I. Kertzer and J. Keith. New York: Cornell University Press.

Scotch, N.A. (1963) Medical Anthropology. In Biennial Review of Anthropology. B.J. Siegel, ed. Stanford: Stanford University Press.

Simic, A. (1985) Ethnicity as a Resource for the Aged: An Anthropological Perspective. Journal of Applied Gerontology 4:65-71

Simmons, L.W. (1945) The Role of the Aged in Primitive Society. New Haven: Yale University Press.

Sokolovsky, J., and C. Cohen (1983) Networks as Adaptation: The Cultural Meaning of Being a "Loner" among the Inner-City Elderly. In Growing Old in Different Societies: Cross-cultural Perspectives. J. Sokolovsky, ed. Belmont, California: Wadsworth Publishing.

Stafford, P. (1982) The Interactional Significance of Senility. Paper presented at the American Anthropological Association meetings. December 5, Washington, D.C.

Subcommittee on Long Term Care of the Special Committee on Aging, U.S. Senate (1975) Nurses in Nursing Homes: The Heavy Burden. Nursing Home Care in the U.S.: Failures in Public Policy, Supporting Paper No. 4, Report NO. 94-00, 94th Congress, 1st Session.

Valle, R. (1989) Cultural and Ethnic Issues in Alzheimer's Disease Family Research. In Alzheimer's Disease Treatment and Family Stress: Directions for Research. E. Light and B. Lebowitz, eds. National Institute of Mental Health. DHS Publications #ADM891569.

Vesperi, M. (1983) The Reluctant Consumer: Nursing Home Residents in the Post-Bergman Era. In Growing Old in Different Societies. J. Sokolovsky, ed. Belmont, California: Wadsworth.

Vicente, L., J. Wiley and R. Carrington (1979) The Risk of Institutionalization Before Death. The Gerontologist 19:361-367.

Virchow, R. 1849 Disease, Life, and Man: Selected Essays. L.J. Rather, ed. Stanford: Stanford University Press.

von Mering, O. (1957) A Family of Elders. In Remotivating the Mental Patient, O. von Mering and S. King eds. New York: Russel Sage Foundation.

von Mering, O (1958) Cultural Values in Normal Senescence, Illness and Death: An Essay in Comparative Gerontology. Psychiatric Communications 1: 63-73.

von Mering, O. (1969) An Anthropmedical Profile of Aging: Retirement from Life into Active Ill Health. Journal of Geriatric Psychiatry 3:61-89.

von Mering, O. and F.L. Weniger (1959) Sociocultural Background of the Aging Individual. In Handbook of Aging and the Individual. J. Birren, Ed. Chicago: University of Chicago Press.

Weiler, G. (1987) The Public Health Impact of Alzheimer's Disease. American Journal of Public Health 77:1157-8.

Wellin, E. (1978) Theoretical Orientations in Medical Anthropology: Change and Continuity Over the Past Half-Century. In Health and the Human Condition. M.H. Logan and E.E. Hunt, eds. North Scituate, Massachusetts: Duxbury Press.

BETHEL ANN POWERS

## 3. NURSING AND AGING

Nursing is a practice discipline whose contributions to studies on aging have tended to be in the form of service oriented research. The review that follows demonstrates the predominance of this focus on problems of care and care delivery and is the basis for observations about the interface of nursing and anthropology within the field of gerontology.

The discussion begins with four concerns that thematically represent the related interests of gerontological nursing and an anthropology of aging. Subsequently, three areas of gerontological nursing research are described with regard to major topics and approaches to subject matter. Particularly illustrative works are highlighted. Closing comments reflect back on earlier expressed interdisciplinary concerns.

### NURSING AND ANTHROPOLOGY

The interface of nursing and anthropology occurs along a boundary marked by historically allied views on (1) a holistic approach to subject matter, (2) attention to participant/client points of view, (3) production of qualitative data through sustained participant observation in natural/clinical settings, and (4) emphasis on normalcy/health and wellness as opposed to deviance/illness (Dougherty and Tripp-Reimer 1985). Because of these similarities, anthropology has had a significant influence on nursing. However, the tendency of anthropologists to misunderstand both the similarities and the differences between the two disciplines has limited the recognition of nursing's actual and potential contributions to anthropology. The root of the misunderstanding is identified by Dougherty and Tripp-Reimer (1985: 219) as a failure to understand the differences between nursing and medicine:

> That nursing is not subsumed by medicine is a point not widely understood in anthropology...The term health care subsumes both nursing and medicine; unfortunately, anthropologists have frequently equated health care only with medicine.

In this paper, the foregoing features of and issues surrounding the interface of nursing and anthropology constitute a point of departure for considering particular concerns in both disciplines that are relevant to gerontology.

*Cross-Cultural Perspectives*

Cross-cultural comparison is the foundation for testing anthropological theory. Principles for unity and diversity in human experience can only be discovered through study of variation. An anthropology of aging must draw on a number of ethnographic cases to uncover patterns and themes that are relevant for a given issue. The goal of the comparative viewpoint in anthropology goes beyond the immediate goals of practice professionals to give culture appropriate care in specific cases, but is, nevertheless, inclusive of them. Therefore, the broader anthropological theories are of use to gerontological nurses in developing research that will support adequate attention to patients' culturally based needs.

There is a growing literature on cultural diversity in nursing practice, developed largely by nurse researchers with anthropology or transcultural nursing backgrounds. As in anthropology, much of the empirical data related to aging and the elderly are embedded in some of those studies. What is apparent in nursing research is a grounding in knowledge of subcultures in the United States. A wide variety of studies using ethnographic methods to describe characteristics of minority cultures has been documented (Tripp-Reimer and Dougherty 1985). They focus on cultural description, health and illness belief systems, professional and lay healers, health behaviors and care practices, and a few have combined cultural and biological variables to study relationships between cultural patterns and natural or disease processes. In addition to broadening the cultural research base, nursing brings to anthropology a distinctive theoretical emphasis on dimensions of caring. This research domain has received little attention from anthropologists. The construct of caring pervades all cultures and extends beyond people's contacts with professional caregivers. The three aspects found cross-culturally are (1) receipt of care, (2) giving of care, and (3) self-care. The combined expression of these aspects of care in cultural context involves a wide range of activities that are often reciprocal in nature and have a role in promoting social cohesion (Dougherty and Tripp-Reimer 1985).

Cross-cultural study of care across the life course is an important path to pursue in gerontological research, where people's needs to develop and maintain self care abilities and achieve some degree of reciprocity in giving and receiving are seen to affect health and well-being. Together, through shared interests combined with distinctive perspectives, nurses and anthropologists could make useful contributions to gerontology in terms of development and application of cross-cultural theory.

*Sociocultural Contextualization*

Anthropologists view age in its totality with particular emphasis on how it is socially constructed and culturally organized. Knowing old people as people demands attention to the environmental contexts from which meaning is derived. A value inherent in an anthropological orientation is the necessity of recognizing diversity. There is a tendency, in research that focuses on individuals, to lose the sense of old people as being very heterogeneous within and across cultures. Systematic attention to sociocultural context prevents the type of agism that leads to treating old people as if they were all alike. Furthermore, the way that anthropologists learn about context, by engaging in the lifeworlds of participants who act as teachers, can work to combat negative stereotyping of elderly as the realities of their situations come to be better understood.

The use of sociocultural contextualization in anthropology and in nursing underscores some differences between the two disciplines. The first cut distinguishes between anthropology as solely a discipline and nursing as both a discipline and a service profession (Dougherty and Tripp-Reimer 1985). The research goals of academic disciplines relate to expanding knowledge through production of theory that is both descriptive and explanatory. In directing their practice, service professions choose among a variety of competing theories and develop prescriptive theories that guide intervention. Therefore, nurses benefit from applying anthropological principles that enhance understanding of their elderly clients' worldviews, health beliefs and practices, and value orientations. However, sociocultural contextualization of client circumstances is a springboard to prescriptive theorizing that can include accepting what is and supporting people as they are or promoting change in those circumstances toward specified practical ends. The first step is consistent with but the latter goes beyond the implied relativism of traditional academic anthropological theory.

The second cut distinguishes between anthropology subdisciplines, variously identified as applied, clinical, or medical anthropology, and clinical nursing. The use of intervention by anthropologists hangs on a number of issues that have to do with standardization and regulation of program content, selection criteria and supervision of students, licensure, certification, practice standards, and approved referral mechanisms. These are all issues that must be addressed by service professions. The direction in which anthropologists in clinical specialty areas may move has not been defined in accordance with these criteria. Their contribution has been in the academic sphere of research that relates to social and clinical service enterprises and policy making. Involvement in actual practice is through collaboration with professionals, such as nurses, who hold social mandates as direct service providers. Gerontology is an area where there are many opportunities for collaboration between nursing and anthropology. Nursing and anthropological approaches to care and aging may be combined and studied within the practice arena to which nursing has access.

*Methodological Diversification*

Nurse researchers use a variety of theoretical perspectives and methods from other disciplines. However, qualitative approaches that have best served anthropology are the least understood. This is beginning to change as nurses employ the full range of qualitative methodologies: ethnography, grounded theory, phenomenology, historical research, and philosophical analyses, in addition to empirical analytic research designs and critical social theory approaches. While not constituting method per se, certain skills and perspectives that come out of nursing subculture and clinical field experience support the rising use and appreciation of qualitative approaches. It is clear that nurses, like anthropologists, have a natural affinity for trying to understand situations from the client point of view. This is routinely evidenced by the appearance of first person accounts and case examples in the non-research literature that reveal use of a discovery mode of inquiry, inductive reasoning, personal reflection, and intuiting client needs (Becker 1983; Cheek 1981; Feigen 1983; Godfrey 1981; Hepler 1982; Huffman 1983; Jenkins 1981; Kenworthy 1983; Klein 1980; Schwartz 1982; Secrest 1984; Timan 1982; Wilson 1981; White 1983).

Nurses' abilities to generate qualitative data come from close and prolonged contact with clients and their families. Often this takes place in or extends into their natural environments: the home and the community. The "caring for" and "being with" emphasis of nurse-client/family interaction is very similar to the participant observation approach used by anthropologists. Nurses have access to clinical settings and some kinds of data less obtainable by social scientists and other health service providers. However, the unique perspectives and depth of knowledge that nurses acquire "in the field" are not always shared outside of the discipline and are often taken for granted by nurses. There are good examples of thoughtful analysis, such as Golander's (1987) ethnographic probing of the rich alive world hidden "under the guise of passivity" by her informants, and Shomaker's (1979) "dialectical etiquette" in the context of nursing home society. Ongoing use and consideration of anthropological approaches as available research design options will help nurses express and share the phenomenological aspects of their practice more broadly and effectively. They make it possible to deal with experiential and intuitive knowledge in a systematic and scholarly way. At the same time, nurses will continue to use diverse methodological approaches in research. This can be a shared strength for anthropology and nursing in examining topics of mutual interest where both measurement and interpretive issues are involved.

*Philosophy and Rationale*

As the anthropology of aging has matured, the need for pulling together scattered studies and formulating a philosophical and theoretical stance has been recognized (Fry 1980). Some work in that direction has been offered by Keith (1980) on the possibility of a general theory of age differentiation to deal with a widening range of data from different societies, and by Kertzer and Keith (1984), on the use of ethnographic analysis in describing and interpreting life course processes. Major contributions of anthropologists have been ethnological descriptions of what it is like to be old in different cultures and analyses of age organization in traditional non-Western societies. In Keith's view, the possibility for interpreting age border issues within a general theory of age differentiation is enhanced and becomes necessary as research extends into complex industrial societies. From a life course perspective (Kertzer and Keith 1984), the opportunity comes in studying aging within cultural systems and across societies as a single evolving process. While historically gerontology has focused on problems of and issues affecting the aged, it is unreasonable to limit aging research to one end of the life spectrum. Increasingly researchers in a variety of disciplines are realizing the importance of placing gerontological studies within a larger framework of social change across time and individual growth and development. Suggestions advanced in these related prescriptions for development of the anthropology of aging aim to situate the unique contributions of anthropology within the matrix of interdisciplinary interests in gerontology.

Gerontological nursing is in a situation similar to that of anthropology. The field is broad and has produced a variety of studies that focus in different directions with no underlying rationale or philosophical base. Across reviews that discuss the state of the art and offer specific suggestions for future research development (Adams 1986; Basson 1967; Brimmer 1979; Burnside 1985; Cora and Lapierre 1986; Gunter and Miller 1977; Kayser-Jones 1981c; Robinson 1981; Wolanin 1983b), the general consensus seems to be that although a great number of phenomena have been addressed, most remain understudied. Emphasis has often been on solving problems associated with functional disability, descriptions of interactions between caregivers and elderly people in different environments (primarily institutional), reports of demonstration projects, and short clinical case examples that illustrate some point related to health assessment and service delivery. There is a need to balance the focus of investigations with increased attention to wellness and health maintenance issues, natural aging, mental health, community based care, and research on administration and consultation to support existing and innovative nursing and multiprofessional health services to the elderly. Some reviewers also have noted the importance of balancing research on psychosocial needs of clients against clinical studies of specific problems that have direct application to nursing care practices, such as sleep patterns, comfort measures, pain relief, infection control, immobility, falls, and physiological control mechanisms.

Reviewers attribute the uneven development of research in gerontological nursing to underemphasis of gerontological content in nursing curricula; shortage of clinically prepared specialists; lack of a critical mass of doctorally prepared researchers to develop the scientific research base; an obsession with some topics, such as attitudes, and the neglect of others; and inconsistent use of theoretical frameworks. While these are very valid points, I suggest that the situation is such where more of everything is not enough. What is needed is an encompassing philosophy and rationale to guide research programs. Gerontological nursing is not only broad in scope, it is also complex. It is both a clinical specialty (subdiscipline) within nursing focused on care of the elderly and a discipline within gerontology, an interdisciplinary field that studies the biological, psychological, and social aspects of aging processes in humans and animals. Research to date has focused heavily on care problems that relate to the clinical specialty. However, it would seem that, as in anthropology, development of gerontological nursing, as discipline and subdiscipline, is firmly tied to its ability to grow along with other researchers and professionals involved in gerontological studies. The previous examples of a theory of age differentiation and life course perspective can serve again to suggest possible shifts in guiding philosophy and rationale toward a broader conceptualization of people's changing health concerns. A sharper underlying emphasis on aging and age differences would produce an overall research agenda that could accommodate and organize the diverse research programs of individuals while simultaneously having a greater impact on encouraging improved health outcomes throughout life into old age. Within this context gerontological nursing's specific clinical knowledge about care of the aged could make a unique contribution to gerontology research and theory development.

Joint attention by nurses and anthropologists to issues of life course and age differentiation also might produce a conceptual framework that is compatible with holistic practice perspectives on aging and health. Anthropology can draw on cross-culturally based theory, and nursing has a strong empirical base from which to draw examples of aging in different contexts that could be used to evaluate the utility and relevance of such a framework. By advancing exploration together, the two disciplines would expand the possibilities for study beyond those that either could address alone.

In summary, the anthropology of aging and gerontological nursing interface lies along the same boundaries as their respective disciplines, with thematic concerns both general to those disciplines and specific to gerontology. It involves similarities in worldview and in theoretical and methodological approaches to subject matter. Also, it involves differences in purpose and mandated mission that influence approaches to contextualization of participant/client experience and structure entry, access, and role of the researcher in fieldwork. The two subdisciplines are broad in scope, and ideas exist on how to manage internal diversity through collective attention to encompassing philosophies and underlying rationales for research agendas. This is particularly important for situating disciplines in an interdisciplinary arena like gerontology. It helps to sort out unique contributions and joint efforts of most use to development of subject matter in the field. In terms of anthropology's commitment to cross-cultural comparison, nurses offer a general base of studies of subgroups in the United States and some important theoretical contributions in the area of care and caring (see Dougherty and Tripp-Reimer 1985 and Tripp-Reimer and Dougherty 1985). Methodological diversification is important for both disciplines when dealing with substantive topics and interdisciplinary interests in gerontology. Nursing's addition of a wide range of qualitative approaches to complement empirical analytic designs makes it a strong resource and ally for anthropologists in exploring topics of mutual interest. The access that nurses often have to settings and data less obtainable by other researchers and their different perspectives that derive from the nature of their interaction with individuals and families also are useful in interdisciplinary endeavors.

## GERONTOLOGICAL NURSING RESEARCH

The following descriptive overview of gerontological nursing literature identifies what is important to nurses in this field and gives illustrative examples rather than exhaustively chronicling work to date. Most publications cited are research based. Occasionally mentioned commentaries or case examples from clinical practice reflect close interplay between the clinical and research arenas. Coverage also reflects some unevenness that exists in the literature and a low incidence of anthropological influence. Cited works by one medical anthropologist (Mitteness) and nurse-anthropologists (e.g. Barbee, Dougherty, Evaneshko, Kay, Kayser-Jones, Morse, Powers, Rempusheski, Tripp-Reimer) show the possibilities for exchange between the two subdisciplines. Informal exchange through the Association for Aging and Anthropology and across institutional settings exists; but the focus here is on published work.

Approaches used in the literature review included (1) computerized and hand searches of 11 nursing and 6 multidisciplinary gerontology journals from first issue through 1987 (some 1988 citations also are included); (2) use of library card catalogs to identify relevant books; (3) use of citations in collected sources to track material; and (4) personal knowledge of work in progress. The focus, though, is on journal publication, viewed as a major means of communication about research within and across specialty areas. From multidisciplinary journals, only work by nurses or a nurse as first author is included.

A brief description of nursing research and backgrounds of those generating the literature precedes the tracing of research themes. Nursing research differs from biomedical research in terms of emphasis and worldview. Nursing is guided by a metaparadigm that defines the discipline's view of phenomena of interest in diagnosing and treating human responses to actual or potential health problems (American Nurses Association 1980). Persons are seen in terms of the attributes of wholeness and integrity and are presumed to be in a continuous state of dynamic interaction with the environment (physical, social, and cultural). Health is a relative concept and a matter of individual perspective that includes but extends beyond physiological status.

The focus on health is in contrast to the medical model focus on disease. Biomedical concerns with care of people who are experiencing pathological changes comprise an aspect of practice that is shared with medicine. Independent nursing practice is concerned with prevention of pathology and helping people to achieve and maintain their highest potential for wellness throughout life. In the care of older people, the situation determines what models will influence the role of the nurse. If disease is the major problem, medical care needs will be emphasized. However, when cure and treatment of disease are not the immediate focus, other considerations will influence nursing judgments about care that is offered. Emphasis may be on providing physical comfort and emotional support to the terminally or chronically ill; on assessing functional abilities and teaching self care skills to people with potential for rehabilitation; or on helping the well elderly adjust to life changes and maintain their health through teaching and assessment of physical, psychological, and social well-being. Because the majority of care and support of the elderly is provided by relatives and friends, an additional focus has been on understanding natural support systems with the aim to preserve and enhance the effectiveness of families as groups and individuals through listening, counseling, teaching, and encouraging.

Although nurses who practice in hospitals or nursing home settings tend to be the most visible to the public, the emphasis in gerontological nursing is consistent with the increasing national focus on promoting home and community based services. In the community gerontological nurses function independently, as members of interdisciplinary teams, or as case managers and directors of organized community service systems.

There is marked internal diversity within nursing in terms of clinical specialization and educational preparation. There are several types of education designed primarily to prepare nurses to practice at the technical level. University preparation at the baccalaureate level is the first step for advancement toward a clinical subspecialty and experience in research. The gerontological nursing literature is generated by nurses who have primary or dual roles as practitioner, administrator, academician, or researcher. Authors' clinical subspecialties may include adult care, rehabilitation, psychiatric mental health, women's health, and community health. Advanced educational backgrounds include clinical specialization at the master's level and preparation in research at the doctoral level in nursing or a related discipline, including anthropology.

## RESEARCH THEMES

Three major categories encompass much of the work done to date: physical health and safety, person-environment interaction, and attitudes and educational priorities. These general categories include different strands of research based on a variety of conceptual and analytic approaches.

## Physical Health and Safety

Nurses have directed a good deal of attention toward patient conditions that present regularly as problems in giving care to the elderly. Major care concerns are confusion, use of medication, urinary incontinence, falls and pressure sores. Other topics that have received less study include sleep patterns, movement and exercise, temperature regulation, bowel care, pain, sensory deficits, diet and nutrition, oral health, and comfort care. Few of these are exclusively problems of the elderly; but from a gerontological viewpoint, what matters is how the particular responses of people in this age group condition the approaches to care.

## Confusion

Most nurse authors writing on confusion note the absence of adequate terminology and the multitude of existing terms that are used interchangeably and inconsistently. The major clinical issue in regard to accurate classification is in distinguishing between confusional states that result from acute reversible processes and those considered to be chronic and irreversible in nature. Physicians may need to make these kinds of diagnoses for patients of any age. The age specific issue is the historical tendency to be less aggressive in diagnosing and treating confusion in the elderly. Nursing recognizes that confusion is not a feature of normal aging but a manifestation of illness. Elderly people are presumed to be more vulnerable to altered cerebral function because of physiological changes, presence of chronic diseases, and sensory disturbances. The literature advocates for increased awareness of the multiple physiological, psychological, and environmental etiologies of this phenomenon and cautions against premature labeling of troubled behavior that might lead to stereotyping (Foreman 1986; Wolanin 1983a). However, there has been little direct study of the needs of cognitively impaired elderly per se. The investigative focus has been on nursing care needs to assess mental status and prescribe appropriate interventions.

Surveys of the assessment and reporting skills of nursing personnel in institutional settings indicate that there are knowledge gaps and variations in the ways that individual care providers define confusion (Brady 1987; Lincoln 1984; Palmateer and McCartney 1985). This may reflect differences in educational backgrounds as well as the evolving state of gerontological knowledge relative to conceptualization of mental confusion and refinement of measurement approaches. Formal assessment in clinical and research situations often involves multiple methods to tap into different aspects of confusion, such as memory, orientation, mood, communication skills, judgment, and emotional control. It is likely, however, that independent measurement scales fail to capture the whole phenomenon. Nagley (1986) suggests that confusion is probably best studied through daily or continuous observation in addition to mental status testing.

Intervention studies focus on nursing behaviors. Williams *et al.* (1979, 1985a) in their study of elderly people hospitalized with hip fractures identify predictors of acute confusional states and the measures that nurses tend to take in response, such as orienting and reassuring patients, and providing explanations. They report significant reduction of confusion when these interpersonal interventions are selectively applied. The quasi-experimental approach yields inconsistent results across studies of this sort, however, because it is not always known how nurses in control or comparison groups are interacting with patients; and reliance on data collection instruments that do not necessarily involve researcher observation and descriptions make it difficult to know the extent to which interventions in experimental groups are carried out (Nagley 1986). The fuller descriptions of interpersonal support of patients by nurses tend to be found in short focused case examples (Ricci 1983). Other studied interventions in institutional settings involve increased stimulation of patients and environmental support. There has been a limited amount of evaluation of group therapies that make use of methods associated with life review, reality orientation and reminiscence. The value of these different experiences with varying results across settings is difficult to establish. A pilot study by a nurse-psychologist team in Sweden, however, is interesting because of its demonstration that a psychodynamic approach may be better than reality orientation in uncovering the memories, thoughts and feelings of severely demented people (Akerlund and Norberg 1986). There is also a recent interest in studying the problem of wandering, in terms of identifying personal characteristics of those at risk and establishing environmental controls to provide safety with the least limitation of freedom (Dawson and Reid 1987; Rader, Doan and Schwab 1985; Rader 1987; Schwab, Rader and Doan 1985).

Some comparative cross-cultural work relates to ethical and quality of life concerns in the force feeding of severely demented institutionalized patients. Swedish studies describe the conflict nurses may encounter when they wish to feed patients who refuse to eat or drink without use of force. Feeding difficulties were found to be reduced when feeders were more consistently assigned to the same patients (Athlin and Norberg 1987; Backstrom, Norberg, and Norberg 1987; Michaelson, Norberg, and Norberg 1987). In a study that used an ethical decision making model to analyze Israeli care providers' thoughts related to force feeding, differences were found that contrasted with the Swedish studies. Jewish sanctity of life ethics were cited as the basis for decreased guilt experienced by caregivers in force feeding demented patients to preserve their lives. A degree of upset about the situation was evident, however, and some caregivers suggested family home care as the alternative where a case by case ethical approach could be applied (Norberg and Hirschfeld 1987).

More diverse and unassembled works in the nursing literature on confusion explore the dimensions of family and professional caregiver experience with mentally impaired older adults for appropriate insights and directions. The impact of confusion on quality of care is suggested in a combined questionnaire and observational study of nurse-patient interaction in a Canadian geriatric ward (Armstrong-Ester and Browne 1986). It was reported that confused patients were more dependent and inactive and that nurses directed less interaction and encouragement of self care activities toward them in contrast to patients who were lucid. Correspondingly, the nurses, when surveyed, identified physical care as their main priority in meeting the needs of confused patients. The challenge to custodial approaches to elderly confused patients that do not rob them of their selfhood lies secondarily in modifying their behavior and primarily with altering family and professional caregivers' interpretations of and reactions to it. The methods used to raise awareness of this need include formal study aimed specifically at understanding patients' problematic behavior (Shomaker 1987); summarization of clinically derived knowledge of interpersonal and environmental techniques to be applied in ways that neither overestimate people's abilities nor underestimate their potential (Bartol 1983); and reflection on care placement of confused people and changes needed in practice and public policy (Lederer 1983; Salisbury and Goehner 1983). A great deal of work needs to be focused on assisting families, caregivers, and policy makers to learn new ways of understanding how elderly "confused" people attempt to cope with life and to increase tolerance of behavior which might ordinarily be resented or found unacceptable.

*Medications*

Nursing studies in this area emphasize drug safety, individualized approaches to drug therapy, and reduced reliance on drugs whenever possible. The drug safety literature focuses attention on both problems of patients failing to comply with prescribed regimens as well as professionals' need to understand and avoid potential drug interactions in cases where people are taking multiple medications. People who are older, isolated, and/or living alone have been found to be least likely to comply with prescribed drug treatments. Errors have been linked to inaccurate knowledge of the drug and its purpose, omission through forgetfulness or confusion over directions, self medication with other non-prescription drugs, and omission because of concerns about side effects. Potential for making the most serious errors rises with the number of medications prescribed and/or absence of a planned method for taking them. Research studies record these dimensions of noncompliance and include nursing implications that focus on approaches to helping people see the relevance of their drugs and understand how to take them responsibly (Alfano 1982; Neely and Patrick 1968; Wade and Bowling 1986).

The responsibility of nurses for administering drugs and monitoring patients for untoward effects underscores the need for knowledge about drug interactions (i.e. modification of the action of one drug by another drug). Studies on this topic are not numerous and have concentrated on institutionalized populations. Computer screening of individual drug profiles has confirmed cases with the potential for clinically significant drug-drug interactions. Potential rises with the number of drugs prescribed. Diuretics and cardiotonics have been found to interact the most followed by analgesics, tranquilizers, and antidepressants (Brown *et al.* 1977; Foxall 1982). Because the elderly react differently than younger people to single or multiple drugs (Hayter 1981), at risk cases have been found even when the person is under the care of a single physician (Burk 1982).

Another approach to dealing with vulnerability of elderly to chemical substances is to try to reduce their dependency as much as possible. Wolanin (1981) asserts that drug therapy should complement nursing care and not be an alternative to it. Physical comfort measures, relaxation techniques, and human interaction also contribute to relief of discomfort and anxiety. Findings of increased symptomatology of patients on certain types of drugs call for more controlled studies and more behavioral interventions as alternatives to drug therapy (Butler, Burgio and Engel 1987); and reports of limited demonstration projects to withdraw specific medications from the regimen of selected patients suggest that there is an absence of ill effects with potential for increased alertness and improved functional abilities (Chisholm, Lundin and Wood 1983; Keenan *et al.* 1983).

Individualized approaches to drug therapy are less frequently reported but indicate potential areas for study of administration of analgesics (Faherty and Grier 1984); timing of sleep medication (Dittmar and Dulski 1977), and self medication for some institutionalized elderly (Meguerdichian 1983). The basis for investigation is the need to understand the interrelationships between chemical effects of drugs, social behavior, and functional abilities.

*Urinary Incontinence*

Many nursing studies on urinary incontinence are intervention studies involving chronically ill, institutionalized patients. Some sort of "bladder training" or "retraining" has been the behavioral approach most often used (Carpenter and Simon 1960; Long 1985). It typically consists of scheduled times for urinating that may be fixed, progressively extended to encourage more control of the urge to urinate, or adapted over time to the person's own pattern of toileting habits. It may also consist of scheduled opportunities for patients who are offered assistance in toileting at timed intervals and are taken to the toilet if they wish. Fewer studies apply concepts of operant conditioning rewarding successful toileting/continent behavior) but they tend also to employ some type of scheduling (Grosicki 1968). Although outcomes of interventions are difficult to measure and study samples tend to be small, some degree of success is usually reported. The greatest difficulty is in determining if the intervention truly changes the patient's behavior in the long run or if it is the behaviors of staff or those responsible for the patient's care at home that have been changed by virtue of instituting the schedule. What is needed is long-term follow-up data on the effectiveness of these kinds of approaches. Other types of studies reported involve clinical trials of products, such as bed pads or diapers, which may be used in care of incontinent people (Grant 1982).

The focus on intervention with institutionalized populations, where urinary incontinence is linked with physical and mental disability, imposes some limitations on understanding the full range of physical, psychological, and social factors that contribute to continence vs. incontinence in noninstitutionalized people. A few nurses have addressed the broader issues, particularly those that account for hiding the symptom of incontinence because of fear, shame, or the belief that it is a normal part of aging (Brink 1980; Simons 1985; Wells 1984) and those that explain the psychological impact on both patients and caregivers in terms of stress (Yu 1987; Yu and Kaltreider 1987). The work of Mitteness (1987), an anthropologist, further confirms the lengths to which elderly people will go to conceal problems of incontinence and underscores the importance of better conceptualization of the phenomenon as a whole.

A recent trend is to give more attention to community based elderly and to develop interventions that are consistent with concepts of preventative treatment and self care (Robb 1985; Taylor and Henderson 1986). There is particular interest in assessing the characteristics of the circumvaginal (pelvic floor) muscles that are thought to be important in maintenance of urinary continence in women. Research efforts involve investigation of the extent to which individuals may be trained (with and without accompanying medication regimes) to implement neuromuscular control mechanisms. Self help approaches must also take into account people's subjective state (perception of incontinence, motivation) and lifestyle (Brink, Wells and Diokno 1987; Dougherty, Abrams and McKay 1986; McKey and Dougherty 1986; Wells, Brink and Diokno 1987).

*Falls*

Falls pose a significant threat to old people in terms of injury and personal expense (e.g. fear and insecurity, decreased mobility, and increased dependence on others). Incidents of falling are, likewise, costly to health care systems, requiring follow-up and redress in terms of treatment and prevention of injuries. Institutional incident report forms have been used extensively to examine epidemiological patterns in fall rates and time of falls, who fallers were, and injuries sustained. Much nursing research emphasizes characteristics of fall-prone elderly derived from observation and retrospective review of patient records (Clark 1985; Craven and Bruno 1986; Venglarik and Adams 1985). Characteristics cover physical and mental condition, mobility and functional capacities, and disease or medical treatment factors that could influence strength, balance, and stability. Precautions are instituted on the basis of the presence of risk factors, with studied results suggesting some combined approaches to be effective in reducing the rate of falls (Daley and Goldman 1987; Hernandez and Miller 1986). Most approaches advocate avoidance of physical or chemical restraint; combine orientation, communication, and supervision techniques with environmental safety control features; and may involve educational programs to increase staff awareness and accountability. In terms of practice applications, though, the knowledge base is restricted. Research on predictor variables is limited by the absence of both controlled studies that compare fallers with nonfallers and descriptive detail concerning the circumstances preceding and accompanying the fall incidents. Only two studies have examined differences between fallers and a randomly selected control group of nonfallers (Lund and Sheafor 1985; Morse, Tylko and Dixon 1985, 1987). Morse's work included qualitative observational and interview data that amplified other reported findings. Both studies confirmed in some respects and contradicted in others the findings of previous studies. The need remains to compare and describe in repeated studies in order to improve current measures for identifying individuals at risk of falling and to take a fuller range of variables into account.

*Pressure Sores*

Treatment and prevention of pressure sores has been a research focus in nursing for over two decades. Particularly in elderly bed or chairbound people, they develop very quickly, persist, and tend to heal very slowly. Because their presence is immediately associated with poor quality of nursing care, occurrence of skin breakdowns make conscientious nurses feel guilty, frustrated, and confused. Most crucial to prevention is identifying those at risk. Ordering of risk factors in terms of frequency of findings vary from study to study, but the ones reported by Pajk *et al.* (1986) are commonly cited: altered nutritional status, impaired activity, impaired mobility, incontinence, and altered mental states (see also Williams 1972). Increased susceptibility to pressure sores with advancing age has been explained as a result of normal skin changes over time, i.e. loss of elasticity and subcutaneous fat, decreased glandular secretions, and generalized skin atrophy. However, in the report of their study of variables associated with skin dryness in the elderly, Frantz and Kinney (1986) suggest that there is much taken for granted knowledge in this area and that some views on the etiology of skin changes and conditions are not tested or well substantiated. It would be appropriate for nurses to do physiological studies to complement continuing research on guides to assess people most at risk of developing pressure sores (Goldstone and Roberts 1980; Gosnell 1973; Norton, McLaren and Exton-Smith 1962; Verhonik 1961) and clinical trials of products and treatment protocols (Boykin and Winland-Brown 1986; Goldstone *et al.* 1982; Jones and Millman 1986; Whitney, Fellows and

Larson 1984). Yet to be studied is the personal impact of skin ulcers from the annoyingly persistent to the overwhelmingly severe.

*Other Research Topics*

There is a recent research interest in sleep patterns that could be important in the nursing care and counseling of elderly people. Colling (1983) has suggested that although the empirical evidence suggests that the elderly experience more sleep disturbances than persons of other age groups, some changes in sleep-wake patterns could be physiologically beneficial and adaptive. Interference with individuals' natural cycles by use of drugs or other measures may be unwarranted and possibly harmful. Hayter (1983, 1985) reported findings consistent with the small number of cross-disciplinary studies in this area, i.e. that variability in sleep behavior increases with age. Her research describes some of the variation. One example deals with daytime napping, which she advises is nothing to be concerned about. It is not an attempt to compensate for sleep lost during the night, and it does not affect nighttime sleep. A comparison of nappers and non-nappers suggested that changes toward increased time spent napping and time awake during the night occur normally in old age and are independent of each other. Different people have different needs. Increased knowledge of differences in sleep requirements has led to investigation of the use of sleep histories to assess individual patterns and research aimed at reevaluation of traditional interventions to control sleep habits through behavior modification and drugs (Clapin-French 1986; Johnson 1985). Briefly, other research topics that have been tentatively explored but not developed specifically in relation to elder care include: movement/activity/exercise in relation to recovery and health (Bassett, McClintock and Schmelzer 1982; Goldberg and Fitzpatrick 1980; Hamilton-Word, Smith and Jessup 1982; Lukens 1986); regulation of body temperature (Higgens 1983; Kolanowski 1981; White *et al.* 1987; Wirtz 1987); treatment of constipation and management of bowel function (Behm 1985; Miller 1985; Pollman, Morris and Rose 1978); pain control (Pearson 1987); tactile sensitivity and taste (McBride and Mistretta, 1986; Mistretta and Oakley 1986; Moore, Nielson and Mistretta 1982; Thornbury and Mistretta 1981); oral health (DeWalt 1975; Engle *et al.* 1985); effects of diagnostic studies on the aged (Robinson and Demuth 1985), and comfort care (Fakouri and Jones 1987; Wagnild and Manning 1985).

PERSON-ENVIRONMENT INTERACTION

Nurses frequently frame their view of practice around issues that arise out of person-environment interaction. In this view, nurses act on behalf of patients to assist with self care that they would perform for themselves if they were able; to bridge formal and informal care systems; and to enable people's continued participation in the everyday worlds that are real and meaningful to them to the greatest extent possible. The dominant research focus has been on nursing care needs. However, increasing attention has been directed toward investigation of person-centered concerns such as elders' perceptions of time; issues of control, adjustment, and support; and social and cultural considerations.

*Nursing Care Needs*

Many studies in the gerontological nursing literature are concerned with identifying needs for nursing care and evaluating its effectiveness. Some examine care needs in relation to nurses' workloads in institutional settings. They attempt to determine ways to group people according to level of need for care and to accurately estimate extent of elderly people's dependence on nursing assistance with regard to such things as feeding, toileting, bathing, dressing, skin care, ambulation, and surveillance (Magid and Hearn 1981; Wade and Snaith 1981). It is at this level that the practical issues of care are addressed. At another level, research addresses political issues that involve such things as the criteria that determine workload patterns and how these relate to meeting needs in long term care. For example, Smith and Molzahn-Scott (1986) compared nursing care requirements in Canadian geriatric long term and acute care settings and asserted that the nursing care needs of the long term care patients were different from but as extensive as needs of patients receiving short term and more complex intensive therapy. They called for better ways to measure nursing workloads and questioned the widely held assumption that institutionalized elderly require less total care than patients receiving more active technologically based treatment. A related issue yet to be addressed is the match between nursing care needs of institutionalized elders and the actual knowledge and skills of their caregivers. The high ratio of nonprofessional to professional nursing staff in long term care is often justified by the belief that care needs are fewer and that the care itself is simpler to perform. This belief promotes hiring less skilled and training unskilled workers. It is reinforced by asserting that professional nurses do not want to give this kind of care. However, there is growing understanding within nursing of the ways in which the custodial approach to care of the elderly in this country has been socially produced. The controlled presence of professional nursing in long term care settings seems also related to economic and political issues within the health care industry. Unless nurses develop objective data bases to systematically document patients' needs and challenge existing standards for allocating nursing resources, elderly patients may remain second class citizens within organized service systems. It is clear that nursing care goals for geriatric clients go beyond physical care to include psychologically and socially based approaches that encourage wellness and personal autonomy in a variety of settings (Barbaro and Noyes 1984; Bergman and Golander 1982; Davies and Crisp 1980; Hallal 1985; Hollinger 1986; Huber and Miller 1984; Hugo *et al.* 1985; Lappe 1987; Shannon 1976). However, many care studies that attempt to address what is actually therapeutic about nursing interventions (Hoch 1987; Kitson 1986) tend to focus on global models of individual adaptation or reactions to stress and on the nature of nurse-client interactions. Social context is either a given or is treated in terms of how it affects individual care outcomes. These kinds of studies have important implications for improving care delivery. But by concentrating more attention on daily practice interactions that make a difference at the individual level, nursing has not developed a critically based literature. Such a literature would focus on social change that could make a difference in terms of broad service system care outcomes.

An important research area that needs to be developed along interdisciplinary lines is that of ethics in care of the elderly. Research indicates variation in nursing actions and motivation for actions in response to care needs that involve such things as decisions to restrain patients (Yarmesch and Sheafor 1984) and aggressive vs. do-not-resuscitate care approaches to older people (Shelley *et al.* 1987). There is also a recognized need for approaches to understand and deal with elder abuse (Fulmer and Cahill 1984; Gilbert 1986; Phillips 1983). The dynamics of decision making that includes client input is often discussed through use of case examples that raise questions about patients' rights to determine their own care and describe resolution of conflicts between nurse and patient care goals (Kennedy 1985; LaGioia 1986). When determining how professional care may affect quality of life for an individual, that person's unique priorities are important factors. The complexities involved in search for ethical principles governing care of people judged to be *not* competent to articulate their own needs is of particular concern. For example, Hirschfeld (1985) argues that neither "sanctity of life" nor "quality of life" positions are acceptable moral guidelines in making care decisions affecting demented older adults. She suggests two ways out of the dilemma: one through more research based clinical knowledge, and the other through a worldview of interdependency combined with a tradition of practical reasoning that draws on Jewish ethical perspectives of life.

Nursing care needs of community based elderly is a research area that focuses attention in two directions. Some studies examine the direct role of nursing in providing services and maintaining continuity of care through such means as discharge planning, home visits, and health programs for the well elderly (Hewner 1986; King, Figge and Harmon 1986; Sullivan and Armignacco 1979; Thornbury and Martin 1983; Waters 1987a). Other studies contribute to the growing cross-disciplinary literature on the dynamics within families involved in caring for elderly members. Some combine a survey of the stresses and limitations imposed by deteriorating health and chronic disorders of elderly persons with suggestions for how nurses should support family caregivers and assess their health needs (Hawranick 1985; Sexton 1984). Lund, Feinhauer and Miller (1985), additionally, reported dissatisfactions at every generational level when grandparents, parents and children live together that suggests the need to think more of families as interactive wholes when considering care needs.

Toward this end, two studies used grounded theory methods to explain what determines patterning and quality of family caregiving (Bowers 1987; Phillips and Rempusheski 1986). This sort of approach to theory development based on lived experience of study participants provides a needed contrast to evaluation measures that draw more attention to the task performance aspects of caregiving. Understanding the family's situation from a variety of perspectives is a priority in practice. Nurses' abilities to assess family involvement in care of older people is essential for identifying and helping to ameliorate high-risk caregiving situations.

The elderly person's circumstances determine the level at which nursing care needs are assessed. In supportive and restorative nursing care the concern is mainly to meet dependency needs that result from a disease state or people's physical condition. Some nursing studies, however, have examined the potential for environmental and social production of dependency in care settings. There has been expressed concern that rigid institutional routines and inadvertent reinforcement of dependent behavior through language and action of nursing staff may contribute to lessened rehabilitation potential and deterioration of elderly in hospitals and nursing homes (Lanceley 1985; Miller 1984; Ryden 1985). Studies also have linked nurses' positive attitudes toward elderly with their belief in ill and institutionalized elders' potential for rehabilitation (Heller, Bausell and Ninos 1984). It seems as though the critical turn from dependent care toward nursing care that stresses health promotion and maintenance rests on recognition of when dependency situations become self-reinforcing. Demonstration projects attempt to show how nurses can prevent this from happening by changing their style of care from a custodial to a rehabilitation focus (Stevenson and Gray 1981). This requires placing attention on what the individual can do instead of on the tasks that the nurse performs for that individual.

Teaching needs of elderly people to support independence in self care have been studied in relation to elders' priorities, motivations, and self care practices (Brown and McCreedy 1986; Cox 1986; Cox, Miller and Mull 1987; Lashley 1987; Smith *et al.* 1980). In this area of nursing research attention is on self actualization of elderly people in all types of settings. Lantz (1985) studied conditions most favorable for fostering achievement of elders' self care goals. He attempted to provide a basis for criteria that could assist nurses in determining the "teachable moment", i.e. that time when the individual is physically, psychologically, and socially most receptive to new learning experiences. Kim (1986) also has examined teaching needs of the elderly in terms of response time and pacing of instruction. Factors identified in other studies that affect the quality and outcome of teaching-learning situations with elders relate to praise, knowledge content, time and attention, and client need for perceived control over health and self care practices (Burckhardt 1987; Harper 1984; Pease 1985; Sands and Holman 1985).

## Person Centered Concerns

The concerns of people who are actual or potential receivers of nursing care have been examined in a variety of ways. Research has tended to focus attention on elderly people's perceptions, morale, and self-esteem. These themes are reflected in work related to (1) perceptions of time, (2) sense of control and personal satisfaction, (3) factors affecting adjustment to life situations, and (4) use of environmental resources for individual support and empowerment.

### Time

There are a small number of studies on time perception, beginning with Newman's (1982) exploration of subjective time as a developmental phenomenon of people's expanding consciousness. The tentative conclusions from this and a later study (Newman and Gaudiano 1984) suggested that perhaps in normal aging one experiences an expanding sense of time and an evolving sense of oneness with the flow of the universe. However, certain mood states, such as depression, may decrease subjective time experience. Practice implications involve finding ways to help old people expand their experience of time and, thereby, improve the quality of their lives. Reminiscence was given as an example of an expanded consciousness state that may represent a significant developmental task of aging. This view of successful aging raises serious questions about interventions with the elderly that concentrate only on increased *inter*activity with others. It supports giving equal attention to measures that facilitate *intra*activity as a means of self actualization. Strumpf (1986, 1987) has reported women's responses to several scales and interview questions about their perceptions of the movement of time and its impact on their lives. Findings suggested that a sense of timelessness may be an important factor in health and well-being for those advanced in years. It was suggested that exploration of the meaning of temporality might give gerontologists and health professionals new insights on physical and mental health and perhaps even longevity. Lastly, Mentzer and Schorr (1986) studied control as a possible factor in conflicting findings about perceived duration of time among the elderly. While perceived duration of time was not found to be significantly related to age or perceived control, length of institutionalization *was* positively related to perceived control. It was suggested that nurses try to find ways to help the institutionalized elderly gain a sense of control in that situation. The gerontological literature on the experience of time is scant. It would seem, though, that further exploration could yield valuable insights with regard to people's adjustment and satisfaction with changing personal circumstances across the life span.

### Control

The issue of control is often linked with personal satisfaction in the literature. Nurse researchers have both assumed and tried to demonstrate that elderly people's morale and self-esteem are higher when they have a greater sense of control over their daily activities (Chang 1978, 1979; Pohl and Fuller 1980; Ryden 1984; Taft 1985). Researchers point out the need for discovering how to increase appropriate options for elders in all settings to have choices and to make their own decisions. It has been emphasized that institutionalization imposes the greatest constraints on elderly people's actual and perceived control of their lives. Few studies, however, explore the social dimensions of this problem. Some support for moving in that direction is found in a study by Slimmer *et al.* (1987), who asked nurses to describe situations that they thought illustrated learned helplessness on the part of elderly patients. All of the examples submitted involved a request for help with a task that nurses thought the person was physically capable of performing. In addition to perceiving these patient behaviors as undesirable, dependent, passive, and rigid, it was believed that they represented an attempt by patients to control others. The researchers, in attempting to generate an operational definition of learned helplessness, were expecting responses more in line with the premise that this condition occurs due to perceived loss of control. The unexpected responses of the nurses suggest that there is a strong need to understand the subcultural influences that regulate social exchange in different settings. Otherwise, the knowledge of how more or less sense of control affects the individual is useless, and appeals to change environments in order to take issues of power and control into account lead nowhere.

*Adjustment*

A variety of factors thought to affect adjustment to changing life situations have been examined. Though there are some studies that suggest that the difficulty and magnitude of change is perceived as being greater by older as opposed to younger adults (Muhlenkamp, Gress and Flood 1975), the results of much research with noninstitutionalized populations supports a view of the great resilience of people who have survived into old age (Edsall and Miller 1978; Fuller and Larson 1980; Melanson and Downe-Wamboldt 1987). This would, then, seem to be a good foundation from which to launch more investigations concerning strategies for successful aging. Abstract notions of what it means to grow old exist in abundance, but some basic conditions that have an impact on meaning at the individual level are understudied. Examples include control of personal possessions (McCracken 1987), dress and appearance (Pensiero and Adams 1987), exercise and physical activity (Parent and Whall 1984), and sexuality (Brower and Tanner 1979; Damrosch 1982; Friedeman 1979). Developmental life process is used by some nurse researchers as a framework within which people's management of basic life conditions across time and space can be understood. Adult development tends to be presented as a progressive rather than a decremental process. Successful aging within this framework has been described in terms of a serial "trading off" of older for more useful behavior patterns (Reed 1983) and in terms of varying states of consciousness (Schorr 1983). A focus on wellness in studies of adjustment and development across life contributes importantly to the major independent function of nursing that is to promote health maintenance.

*Support*

Understanding how people use environmental resources for support and empowerment is important in assessing need for professional services and intervening appropriately. There is a large body of cross-disciplinary literature on social support to which nursing has contributed. Most of those nursing contributions that deal specifically with the elderly are in the area of bereavement. Studies of mate loss as a stressful life event often focus on the association between grief and health risks (Brock 1984; Gass 1987; Remondet and Hansson 1987; Valanis, Yeaworth and Mullis 1987). Rigdon, Clayton and Dimond (1987) developed a dialectical theory of helpfulness through content analysis of responses given by bereaved people as to the kinds of advice they gave to others in similar circumstances. Much had to do with maintaining and mobilizing personal resources. Crisis theory has also been used to describe imbalance of needs and resources that may accompany mate loss and to discuss the dynamics of nursing interventions (Richter 1984, 1987). Particular consideration has been given to the role of personal network resources in model development to account for variation in adaptation among elderly bereaved (Dimond 1981) and to determine the role of structural and qualitative components of social support in depression, coping, health, and life satisfaction (Dimond, Lund and Caserta 1987). A much smaller sampling of nursing literature on social support can be cited in relation to the health of community based elderly (Ide 1983; Laschinger 1984; Preston and Grimes 1987), social networks of elderly institutionalized people (Powers 1988a), and mental health and depression (Hatcher, Durham and Richey 1985; Reed 1986). Similarly, apart from Robb's work, little research has been done on companion animals (Robb, Boyd and Pristach 1980; Robb 1983; Robb and Stegman 1983) or the effect of dolls and toy animals that are sometimes collected by or given to hospitalized adults (Baly 1986; Milton and MacPhail 1985). An expanding literature, however, deals with women's issues relative to support and life satisfaction (Chang *et al.* 1984; Gelein 1980; Hoeffer 1987; Spotts 1981), health practices (Chang *et al.* 1985; Kolanowski and Gunter 1985), body image (Janelli 1986), and feminist perspectives as alternatives to medical and social metaphors that have to do with female reproductive functions and sexuality, notably menopause and osteoporosis (MacPherson 1981, 1985).

*Social and Cultural Issues*

In the gerontological nursing literature, considerations of social and cultural contexts are often embedded in explorations of client problems and nursing care needs. The extent to which context is featured varies.

Studies of effects of social environment on the lives of elderly people have explored territorial and privacy issues in nursing homes (Johnson 1979; Roosa 1982); placement, relocation, and movement within and between institutions and outside communities (Amenta, Weiner and Amenta 1984; Brock and O'Sullivan 1985; Engle 1985; Grier 1977; Lewis, Messner and McDowell 1985; Petrou and Obenchain 1987; Waters 1987b; Wiltzius, Gambert and Duthie 1981); and needs of rural elderly (Barber et al. 1984; Schwartz 1980).

Increasing attention to cultural perspectives in gerontological nursing practice and research has come about through the work of nurse-anthropologists, some international exchange between nurse educators and researchers, and the recent prevalence of opportunities for nurses to travel and participate in study tours that expose them to life and health care systems in other countries.

Nurse-anthropologists have promoted the importance of culture specific health care in pluralistic societies: such as the United States, as well as the need for cross-cultural comparative study of health beliefs and practices. "Transcultural Nursing Care of the Elderly" was the theme of the second National Transcultural Nursing Conference in 1977. The goal was to present studies of the aged in Western and non-Western societies: aging and the Black diaspora in terms of African, Caribbean, and Afro-American experience (Osborne 1977); Italian-American (Ragucci 1977); Filipino (Shimamoto 1977); Mexican-American (Benavidez-Clayton 1977); and Mormon (Peay 1977). While one paper on values, beliefs and practices of elderly women in the United States (Sullivan 1977) advanced the notion of the aged as a subculture, other nurse-anthropologists have noted the problem with lumping all elderly together without regard for their cultural heterogeneity (Leininger 1976; Tripp-Reimer 1980).

Another nurse-anthropologist used an ethnographic cross-cultural comparative approach to search for answers to American problems with long term care of the aged. Kayser-Jones (1979, 1981a, 1981b) analyzed factors affecting nursing home care of elderly in Scotland and the United States and proposed an exchange theory-based explanation of differences observed in the two institutions that were studied. Cultural attitudes and institutional structures that acted as barriers to quality care were identified as bases for less satisfactory service delivery in the American facility.

In a special issue of the *Journal of Cross-Cultural Gerontology*, concepts and research methods of anthropology plus the practice orientation of nursing are blended and reflected in examples of individual and interdisciplinary studies involving community and institution based populations, women's health issues, and cultural diversity. The authors (first authors of jointly authored papers) are nurse-anthropologists who bring perspectives from both disciplines to research on: patterns of elder care in an

Old Order Amish Community (Tripp-Reimer et al. 1988); widows, coping, and health (Kay et al. 1988); aging of middle-aged women (Barbee 1988); the meaning of caring for second generation Polish American elders (Rempusheski 1988); an intergenerational geriatric remotivation program (Hutchinson and Webb 1988); and self-perceived health of elderly institutionalized people (Powers 1988b).

Apart from the studies of nurse-anthropologists, most of the culture related literature in gerontological nursing is not research based. Some standard interviewing and measurement techniques have been used to assess life satisfaction among elderly urban Blacks (Johnson, Cloyd and Wer 1982) and reservation dwelling American Indians in the mid-west (Johnson et al. 1986). However, much of the cultural bases for interpretation of findings is absent. A more detailed study of differences and similarities in self-perceptions of aging across cultures: Anglo-American, Chinese-American, and Taiwanese Chinese, was reported by Tien-Hyatt (1987). Other research reports included a survey of over-the-counter drug use by Chinese and Hispanic Americans, showing different preferences for types of preparations and a blending of folk and modern medicine (Hess 1986); and a Canadian study of communication patterns between nursing staff and three different groups of ethnic elderly (Jones and vanAmelsvoort-Jones 1986).

Commentary on culture and elderly people in the United States has touched on concerns of Black elderly (Bailey and Walker 1982; Clavon 1986; Clavon and Smith 1986); elderly in Appalachia (Lewis, Messner and McDowell 1985; Simon 1987); Mexican Americans (Wilson and Heinert 1977); elderly Armenians (Saunders 1984); older Asians (Chae 1987); institutionalized Filipinos (Caringer 1977); comparative observations/Anglo, Jewish and Black American women (Seabrooks, Kahn and Gero 1987); and problems of the affluent elderly, whom Alford (1978) cautioned are likely to be ignored because it may be believed that since they have money they have no difficulties with health and aging. Additional reports of care programs, practice experiences, and approaches to teaching students about ethnicity and aging have appeared (Baer and Gress 1980, 1983; Davis 1980; Dicharry 1986; Evans 1982; Kowalsky 1980; Safier 1976). The overall effect of these types of articles is one of heightened awareness of variability of aging experiences among different segments of the society.

In terms of international perspective, American nurses have published observations made on study tours abroad. Most frequent reports have been on elder care in the USSR (Benson 1978; Duncan 1982; Gioiella 1983) and China (Brower 1984; Chang 1980; Devine 1980a; Martinson 1982; Morrisy 1983; Turkoski 1985). Observation of British geriatric health care (Grau 1986; Podskalny and Woods 1983) and long term care priorities in the European Economic Community (McCall 1974) have been commented upon. Reports on elder care in Japan (Engel, Kojima and Martinson 1986); Egypt (Abd El Ghany 1986); Micronesia (Shimamoto 1984); and developing countries (Martin 1984) have been published as well as discussions of the role of nursing in shaping health care policy in collaboration with agencies such as the World Health Organization (Hirschfeld 1987; Porter 1984) and the World Assembly on Aging (Fulmer 1982). The work is not extensive nor is it in-depth, but it is indicative of movement and growth.

## ATTITUDES AND EDUCATIONAL PRIORITIES

There has been a major continuing effort to survey attitudes of student nurses and care providers toward the elderly (Campbell 1971; Delora and Moses 1969; Devine 1980b; Downe-Wamboldt and Melanson 1985; Dye 1979; Fisk 1984; Gillis 1973; Gunter 1971; Hart, Freel and Crowell 1976; Kayser and Minnigerode 1975; Knowles and Sarver 1985; Melanson and Downe-Wamboldt 1985; Penner, Kudenia and Mead 1984; Robb 1979; Shimamoto and Rose 1987; Smith, Jepson and Perloff 1982; Snape 1986; Taylor and Harned 1978). Study findings suggest that there is a relationship between a tendency to stereotype elders and willingness to care for them. Understanding factors that affect behavior, such as attitudes, is important in improving care delivery. However, a frequently expressed concern is that research activity in this area has been excessive at the expense of equally important study topics. It is likely that the interest in attitudes is linked to educational priorities. Until there is more of a shift to accommodate gerontological nursing content, the study of attitudes will probably continue. The view taken here is that this is not an overstudied area, but there has been too great a proliferation of studies that are limited in their ability to extend knowledge. There is, therefore, a need to address problematic aspects in research focus and design. There have been few attempts to relate people's reported attitudes to their actual behavior. An exception is a study by Hatton (1977) involving both measurement of attitudes and observations of nurses' interactions with patients. Secondly, there has been little systematic study of the bases for attitudes toward the elderly. It is widely held that gerontological content in nursing school curricula and role modeling by nursing faculty and clinicians are major influencing factors (Chaisson 1980; Eddy 1986; Gomez *et al.* 1985; Hannon 1980; Tobiason, Knudson and Stengel 1979). However, studies that attempt to measure change produced by new educational experiences tend to have methodological flaws, often in the areas of recording baseline measures and establishing adequate controls. Last, the tendency to focus on individual caregiver characteristics de-emphasizes the role that social environments play in shaping attitudes and behavior. There is a need to provide an alternative to psychologically based explanations of individual behavior derived from survey measures and testing.

## DISCUSSION

Interplay between clinical practice and research arenas has shaped the development of gerontological nursing. The earliest and best developed areas of research have been service oriented. Emphasis has been on (1) practical solutions to care problems, such as incontinence and pressure sores, and identifying elderly at risk of developing problems, such as falling or wandering; (2) caregiver knowledge and ability to intervene, such as in mental status assessments, management of drug therapies, and clinical trials of physical care regimens; (3) dependent (supportive/restorative) aspects of care, such as assessment of clients' and families' needs for nursing care, distribution and management of nursing care delivery resources, and supportive aspects of nurse-client interactions; and (4) institutionally based care in acute hospital and nursing home settings. Building on this foundation, psychosocial research has emphasized perceptions, morale, and self-esteem of clients as individual receivers of care. And the most current emphasis is on self care (health promotion and maintenance) and community based services, which brings issues of sociocultural contexts forward.

The movement of knowledge development within gerontological nursing has relevance for anthropologists involved in aging research. *Anthropological perspectives* are especially helpful in community studies and in studies of physical health and safety, where social context is neglected. These latter kinds of problems are not outside the domain of interest or expertise of anthropologists. *Nursing perspectives* on caring and health teaching can be illuminating for anthropologists. Care interventions in nursing practice, conceptually and substantively, differ from medical care. And nursing approaches to cultural diversity and cross-cultural comparison in care of the elderly are very compatible with the interests of anthropologists. *Shared perspectives* on social and cultural issues within a conceptual framework that addresses age differences and aging across the life course has theoretical and practical implications for both nursing and anthropology. A potential area of mutual interest is the biocultural. The links between physiological nursing and physical anthropology have not been well explored.

The time is good for cross-disciplinary exchange. Both subdisciplines are relatively established and expanding. Each has unique contributions to make, but research in aging must not be confined within disciplinary boundaries. Problems originating in one discipline can fuel research efforts in others; and new perspectives can develop that become part of the knowledge pool to which all have access.

## REFERENCES

Abd El Ghany, N.I. (1986) Elderly in Egypt. Journal of Gerontological Nursing 12:35-38.
Adams, M. (1986) Aging: Gerontological Nursing Research. In Annual Review of Nursing Research. H. H. Werley, J. J. Fitzpatrick, R. L. Taunton, eds. (Vol. 4, pp. 77-103). New York: Springer Publishing Co.
Akerlund, B. M. and A. Norberg (1986) Group Psychotherapy With Demented Patients. Geriatric Nursing 7:83-84.
Alfano, G. (1982) Meaning of the Medication: Clue to Acceptance or Rejection. Geriatric Nursing 3:28-30.
Alford, D. M. (1978) The Affluent Elderly: Problems in Nursing Care. Journal of Gerontological Nursing 4:44-47.
Amenta, M., A. Weiner and D. Amenta (1984) Successful Relocation of Elderly Residents. Geriatric Nursing 5:356-360.
American Nurses' Association (1980) Nursing: A Social Policy Statement. Kansas City, MO: American Nurses' Association.
Armstrong-Esther, C. A. and K. D. Browne (1986) The Influence of Elderly Patients' Mental Impairment on Nurse-Patient Interaction. Journal of Advanced Nursing 11:379-387.
Armstrong-Esther, C. A. (1986) The Influence of Elderly Patients' Mental Impairment on Nurse-Patient Interaction. Journal of Advanced Nursing 11:379-387.
Athlin, E. and A. Norberg (1987) Caregivers' Attitudes to and Interpretations of the Behavior of Severely Demented Patients During Feeding in a Patient Assignment Care System. International Journal of Nursing Studies 24:145-153
Backstrom, A., A. Norberg and B. Norberg (1987) Feeding Difficulties in Long-Stay Patients at Nursing Homes. International Journal of Nursing Studies 24:69-76.
Bahr, R. T. and L. D. Gress (1980) Course Description: The Nursing Process: Ethnicity and Aging. Journal of Gerontological Nursing 6:210-213.
Bahr, R. T. and L. D. Gress (1983) Education in Gerontology: A Course Description For Several Facets on Gerontology. Journal of Gerontological Nursing 9:384-397.
Bailey, F. E. and M. L. Walker (1982) Socioeconomic Factors and Their Effects on the Nutrition and Dietary Habits of the Black Aged. Journal of Gerontological Nursing 8:203-207.
Baly, G. F. A. (1986) Plush Animals – Do They Make A Difference? Geriatric Nursing 7:140-142.
Barbaro, E. L. and L. E. Noyes (1984) A Wellness Program for a Life Care Community. The Gerontologist 24: 568-571.
Barbee, E. L. (1988) Feeling Older and Wanting to be Younger. Journal of Cross-Cultural Gerontology 3:209-221.
Barber, H., D. Jelenek, M. Barbe, G. Libo, and J. Randals (1984) Helping the Rural Elderly. Journal of Gerontological Nursing 10:104-109
Bartol, M. A. (1983) Reaching the Patient. Geriatric Nursing 4:234-236.
Bassett, C., E. McClamrock, and M. Schmelzer (1982) A 10-Week Exercise Program for Senior Citizens. Geriatric Nursing 3:103-105.
Basson, P. H. (1967) The Gerontological Nursing Literature Search: Study and Results. Nursing Research 16:267-272.
Becker, M. (1983) In Loving Remembrance. Geriatric Nursing 4:176B-178.
Behm, R. M. (1985) A Special Recipe to Banish Constipation. Geriatric Nursing 6:216-217.
Benavidez-Clayton, C. (1977) Variations and Adaptations in Nursing Care of the Elderly Mexican American. In Proceedings: Transcultural Nursing Care of the Elderly, M. Leininger, ed. Second National Transcultural Nursing Conference.
Benson, E. R. (1978) Observations on Health Care for the Elderly in the USSR. Journal of Gerontological Nursing 4:18-20.
Bergman, R. and H. Golander (1982) Evaluation of Care for the Aged: A Multipurpose Guide. Journal of Advanced Nursing 7:203-210.

Bowers, B. J. (1987) Intergenerational Caregiving: Adult Caregivers and Their Aging Parents. Advances in Nursing Science 9:20-31.
Boykin, A. and J. Winland-Brown (1986) Pressure Sores: Nursing Management. Journal of Gerontological Nursing 12:17-21.
Brady, P. F. (1987) Labeling of Confusion in the Elderly. Journal of Gerontological Nursing 13:29-32.
Brimmer, P. F. (1979) Past, Present and Future in Gerontological Nursing Research. Journal of Gerontological Nursing 5:27-34.
Brink, C. (1980) Promoting Urine Control in Older Adults: Assessing the Problem. Geriatric Nursing 1:241-245.
Brink, C. A., T. J. Wells, and A. C. Diokno (1987) Urinary Incontinence in Community Living Women. Public Health Nursing 4:114-119.
Brock, A. M. and P. O'Sullivan (1985) A Study to Determine What Variables Predict Institutionalization of Elderly People. Journal of Advanced Nursing 10:533-537.
Brower, H. T. and L. A. Tanner (1979) A Study of Older Adults Attending a Program on Human Sexuality. Nursing Research 28:36-39.
Brower, H. T. (1984) A Look at China's Elder Care. Geriatric Nursing 5:250-253.
Brown, J. S. and M. McCreedy (1986) The Hale Elderly: Health Behavior and Its Correlates. Research in Nursing and Health 9:317-329.
Brown, M. M., J. K. Boosinger, M. Henderson, and S. S.Rife. (1977) Drug-Drug Interactions Among Residents in Homes for the Elderly. Nursing Research 26:47-52.
Burckhardt, C. S. (1987) The Effect of Therapy on the Mental Health of the Elderly. Research in Nursing and Health 10:277-285.
Burk, J. (1982) Simon Farber, A Man on a Complex Regimen. Geriatric Nursing 3:41-43.
Burnside, I. (1985) Gerontological Nursing Research: 1975-1984. In Overcoming the Bias of Ageism in Long Term Care. NLN Publication 20-1975. New York: National League for Nursing.
Butler, F. R., L. D. Burgio, and B. T. Engel (1987) Neuroleptics and Behavior: A Comparative Study. Journal of Gerontolgoical Nursing 13:15-19.
Campbell, M. E. (1971) Study of the Attitudes of Nursing Personnel Toward the Geriatric Patient. Nursing Research 20:147-151.
Caringer, B. (1977) Caring for the Institutionalized Filipino. Journal of Gerontological Nursing 3:33-37.
Carpenter, H. A. and R. Simon (1960) The Effect of Several Methods of Training on Long Term, Incontinent, Behaviorally Regressed Hospitalized Psychiatric Patients. Nursing Research 9:17-22.
Chae, M. C. (1987) Older Asians. Journal of Gerontological Nursing 13:10-17.
Chaisson, C. M. (1980) Life Cycle: A Social Simulation Game to Improve Attitudes and Responses to the Elderly. Journal of Gerontological Nursing 6:587-592.
Chaisson, C. M. (1980) Life Cycle: A Social Simulation Game to Improve Attitudes and Responses to the Elderly. Journal of Gerontological Nursing 6:587-592.
Chang, B. L. (1978) Generalized Expectancy, Situational Perception,and Morale Among Institutionalized Aged. Nursing Research 27:316-324.
Chang, B. L. (1979) Locus of Control, Trust, Situational Control and Morale of the Elderly. International Journal of Nursing Studies 16:169-181.
Chang, B. L., G. C. Uman, L. S.Linn, and J. E. Ware (1984) The Effect of Systematically Varying Components of Nursing Care on Satisfaction in Elderly Ambulatory Women. Western Journal of Nursing Research 6:367-386.
Chang, B. L., G. C. Uman, L. S. Linn, J. E.Ware (1985) Adherence to Health Care Regimens Among Elderly Women. Nursing Research 34:27-1.
Chang, M. K. (1980) Care of the aged in the People's Republic of China. Journal of Gerontological Nursing 6: 222-224.

Cheek, M. V. (1981) A Change of Life. Geriatric Nursing 2:192-194.
Chisholm, M., S. Lundin, and J.Wood (1983) Withdrawing Digoxin-Worth a Try. Geriatric Nursing 4:290-292.
Clapin-French, E. (1986) Sleep Patterns of Aged Persons in Long Term Care Facilities. Journal of Advanced Nursing 11:57-66.
Clark, G. A. (1985) A Study of Falls Among Elderly Hospitalized Patients. Australian Journal of Advanced Nursing 2:34-44.
Clavon, A. (1986) The Black Elderly. Journal of Gerontological Nursing 12:6-12.
Clavon, A. M. and V. P. Smith (1986) One Black Couple's Means of Coping. Journal of Gerontological Nursing 12:26-28.
Colling, J. (1983) Sleep Disturbances in Aging: A Theoretic and Empiric Analysis. Advances in Nursing Science 6:36-44.
Cora, V. L., and E. D. Lapierre (1986) ANA Speaks Out: Results Based on a Four-Year Survey of Nursing Literature. Journal of Gerontological Nursing 12:21-26.
Cox, C. L. (1986) The Interaction Model of Client Health Behavior: Application to the Study of Community-Based Elders. Advances in Nursing Science 9:40-57.
Cox, C. L., E. H. Miller, and C. S. Mull (1987) Motivation in Health Behavior: Measurement, Antecedents, and Correlates. Advances in Nursing Science 9:1-15.
Craven, R.and P. Bruno (1986) Teach the Elderly to Prevent Falls. Journal of Gerontological Nursing 12:27-33.
Daly, I. and L. Goldman (1987) A Closer Look At Institutional Accidents. Geriatric Nursing 8:64-67.
Damrosch, S. P. (1982) Nursing Students' Attitudes Toward Sexually Active Older Persons. Nursing Research 31:252-255.
Davies, A. D.and A. G. Crisp (1980) Setting Performance Goals in Geriatric Nursing. Journal of Advanced Nursing 5:381-388.
Davis, R. (1980) Health Screening in Hawaii. Geriatric Nursing 1:187-189.
Dawson, P. and D. W. Reid (1987) Behavioral Dimensions of Patients at Risk of Wandering. The Gerontologist 27:104-107.
Delora, J. R. and D. V. Moses (1969) Specialty Preferences and Characteristics of Nursing Students in Baccalaureate Programs. Nursing Research 18:137-144.
Devine, B. A. (1980a) China's "House of Respect". Journal of Gerontological Nursing 6:338-340.
Devine, B. A. (1980b) Old Age Stereotyping: A Comparison of Staff Attitudes Toward the Aged. Journal of Gerontological Nursing 6:25-31.
DeWalt, E. M. (1975) Effect of Timed Hygienic Measures on Oral Mucosa in a Group of Elderly Subjects. Nursing Research 24:104-108.
Dicharry, E. K. (1986) Delivering Home Health Care to the Elderly in Zuni Pueblo. Journal of Gerontological Nursing 12:25-29.
Dimond, M. (1981) Bereavement and the Elderly: A Critical Review With Implications for Nursing Practice and Research. Journal of Advanced Nursing 6:461-470.
Dimond, M., D. A. Lund, and M. S. Caserta (1987) The Role of Social Support in the First Two Years of Bereavement in an Elderly Sample. The Gerontologist 27:599-604.
Dittmar, S. S. and T. Dulski (1977) Early Evening Administration of Sleep Medication to the Hospitalized Aged: A Consideration in Rehabilitation. Nursing Research 26:299-303.
Dougherty, M. C. and T. Tripp-Reimer (1985) The Interface of Nursing and Anthropology. Annual Review of Anthropology 14:219-241.
Dougherty, M. C., R. Abrams, and P. L. McKey (1986) An Instrument to Assess the Dynamic Characteristics of the Circumvaginal Musculature. Nursing Research 35:202-206.
Downe-Wamboldt, B. L. and P. M. Melanson (1985) A Descriptive Study of the Attitudes of Baccalaureate Student Nurses Toward the Elderly. Journal of Advanced Nursing 10:369-374.
Duncan, L. (1982) Observations of Eldercare in the USSR. Geriatric Nursing 3:257-259.
Dye, C. A. (1979) Attitude Change Among Health Professionals: Implications for Gerontological

Nursing.Journal of Gerontological Nursing 5:30-35.
Eddy, D. M. (1986) Before and After Attitudes Toward Aging in a BSN Program. Journal of Gerontological Nursing 12:30-34.
Edsall, J. O., and L. A. Miller (1978) Relationship Between Loss of Auditory and Visual Acuity and Social Disengagement in an Aged Population. Nursing Research 27:296-298.
Engel, N. S., M. Kojima and I. M. Martinson (1986) A Community Health Center Responds to the Aging of Japan. Journal of Gerontological Nursing 12:12-16.
Engle, V. (1985) Temporary Relocation:Is It Stressful to Your Patient? Journal of Gerontological Nursing 11:28-31.
Engle, V., A. Whall, C. Dimond, and L. Bobel (1985) Tardive Dyskinesia: Are Your Older Clients at Risk? Journal of Gerontological Nursing 11:25-27.
Evans, B. (1982) Intervention vs. Interference. Geriatric Nursing 3:244-247.
Faherty, B. S. and M. R. Grier (1984) Analgesic Medication for Elderly People Post-Surgery. Nursing Research 33:369-372.
Fakouri, C. and P. Jones (1987) Slow Stroke Back Rub. Journal of Gerontological Nursing 13:32-35.
Feigen, J. (1983) Divided Loyalties. Geriatric Nursing 4:298-300.
Fisk, V. R. (1984) When Nurses' Aides Care. Journal of Gerontological Nursing 10:118-127.
Foreman, M. D. (1986) Acute Confusional States in Hospitalized Elderly: A Research Dilemma. Nursing Research 35:34-37.
Foxall, M. J. (1982) Elderly Patients at Risk of Potential Drug Interactions in Long-Term Care Facilities. Western Journal of Nursing Research 4:133-151.
Frantz, R. A. and Kinney, C. K. (1986) Variables Associated With Skin Dryness in the Elderly. Nursing Research 35:98-100.
Friedeman, J. S. (1979) Development of a Sexual Knowledge Inventory for Elderly Persons. Nursing Research 28:372-374.
Fry, C. L. (1980) Aging in Culture and Society: Comparative Viewpoints and Strategies. Brooklyn, New York: J. F. Bergin Publishers.
Fuller, S. S. and S. B. Larson (1980) Life Events, Emotional Support, and Health of Older People. Research in Nursing and Health 3:81-89.
Fulmer, T. (1982) A Special Report: The World Assembly on Aging. Journal of Gerontological Nursing 8:702-705.
Fulmer, T. and V. M. Cahill (1984) Assessing Elder Abuse: A Study. Journal of Gerontological Nursing 10:16-20.
Gass, K. A. (1987) The Health of Conjugally Bereaved Older Widows: The Role of Appraisal, Coping,and Resources. Research in Nursing and Health 10:39-47.
Gelein, J. L. (1980) The Aged American Female: Relationships Between Social Support and Health. Journal of Gerontological Nursing 6:69-73.
Gilbert, D. A. (1986) The Ethics of Mandatory Elder Abuse Reporting Statuses. Advances in Nursing Science 8:51-62.
Gillis, M. (1973) Attitudes of Nursing Personnel Toward the Aged. Nursing Research 22:517-520.
Gioiella, E. C. (1983) Russia: The Soviet Health Care System. Journal of Gerontological Nursing 9:582-585.
Godfrey, J. (1981) John, A Severely Regressed Man. Geriatric Nursing 2:281-282.
Golander, H (1987) Under the Guise of Passivity. Journal of Gerontological Nursing 13:26-31.
Goldberg, W. G.and J.J.Fitzpatrick (1980) Movement Therapy with the Aged. Nursing Research 29:339-346.
Goldstone, L. A. and B. V. Roberts (1980) A Preliminary Discriminant Function Analysis of Elderly Orthopaedic Patients Who Will or Will Not Contract a Pressure Sore. International Journal of Nursing Studies 17:17-23.
Goldstone, L. A., M. Norris, M. O'Reilly, and J. White (1982) A Clinical Trial of a Bead Bed System for the Prevention of Pressure Sores in Elderly Orthopaedic Patients. Journal of

Advanced Nursing 7:545-548.
Gomez, G. E., D. Otto, A. Blattstein, and E. A. Gomez (1985) Beginning Nursing Students Can Change Attitudes About the Aged. Journal of Gerontological Nursing 11:6-11.
Gosnell, D. (1973) An Assessment Tool to Identify Pressure Sores. Nursing Research 22:55-59.
Grant R. (1982) Washable Pads or Disposable Diapers. Geriatric Nursing 3:248-251.
Grau L. (1986) Britain's Community Psychiatric Nursing Teams. Geriatric Nursing 7:143-147.
Grier, M. R. (1977) Choosing Living Arrangements for the Elderly. International Journal of Nursing Studies 14:69-76.
Grosicki, J. P. (1968) Effect of Operant Conditioning on Modification of Incontinence in Neuropsychiatric Geriatric Patients. Nursing Research 17:304-311.
Gunter, L. M. (1971) Student's Attitudes Toward Geriatric Nursing. Nursing Outlook 19:466-469.
Gunter, L. M. and J. C. Miller (1977) Toward a Nursing Gerontology. Nursing Research 26:208-221.
Hallal, J. C. (1985) Nursing Diagnosis:An Essential Step To Quality Care. Journal of Gerontological Nursing 11:35-38.
Hamilton-Word V., F. W. Smith, and E. Jessup (1982) Physical Fitness on a VA Hospital Unit. Geriatric Nursing 3:260-262.
Hannon, J. (1980) Effect of a Course on Aging in a Graduate Nursing Curriculum. Journal of Gerontological Nursing 6:604-614.
Harper, D. C. (1984) Application of Orem's Theoretical Constructs to Self-Care Medication Behaviors in the Elderly. Advances in Nursing Science 6:29-46.
Hart, L., M. I. Freel, and C. M. Crowell (1976) Changing Attitudes Toward the Aged and Interest in Caring for the Aged. Journal of Gerontological Nursing 2:10-16.
Hatcher, B. J., J. D. Durham, and M. Richey (1985) Overcoming Stroke-Related Depression. Journal of Gerontological Nursing 11:34-39.
Hatton, J. (1977) Nurses' Attitude Toward the Aged: Relationship to Nursing Care. Journal of Gerontological Nursing.
Hawranik, P. (1985) Caring for Aging Parents: Divided Allegiances. Journal of Gerontological Nursing 11:19-22.
Hayter, J. (1981) Why Response to Medication Changes with Age. Geriatric Nursing 2:411-416.
Hayter, J. (1983) Sleep Behaviors of Older Persons. Nursing Research 32:242-246.
Hayter, J. (1985) To Nap or Not To Nap? Geriatric Nursing 6:104-106.
Heller, B. R., R. B. Bausell, and M. Ninos (1984) Nurses' Perceptions of Rehabilitation Potential of Institutionalized Aged. Journal of Gerontological Nursing 10:22-25.
Hepler, M. (1982) Facing Death Together. Geriatric Nursing 3:394-395.
Hernandez, M. and J. Miller (1986) How to Reduce Falls. Geriatric Nursing 7:97-102.
Hess, P. (1986) Chinese and Hispanic Elders and OTC Drugs. Geriatric Nursing 7:314-318.
Hewner, S. (1986) Bringing Home the Health Care:Nurses Make a Difference. Journal of Gerontological Nursing 12:29-35.
Higgens, P. (1983) Can 98.6 Be A Fever in Disguise? Geriatric Nursing 4:101-102.
Hirschfeld, M. J. (1985) Ethics and Care for the Elderly. International Journal of Nursing Studies 22:319-328.
Hirschfeld, M. J. (1987) The World Health Organization's Regions of the Eastern Mediterranean and Europe: Ageing of the Population and Nursing Care. Journal of Advanced Nursing 12:151-158.
Hoch, C. C. (1987) Assessing Delivery of Nursing Care. Journal of Gerontological Nursing 13:10-17.
Hoeffer, B. (1987) Predictorts of Life Outlook of Older single Women. Research in Nursing and Health 10:111-117.
Hollinger, L. M. (1986) Communicating with the Elderly. Journal of Gerontological Nursing 12:8-13.
Huber, K. and P. Miller (1984) Reminisce with the Elderly: Do It! Geriatric Nursing 5:84-87.
Huffman, J. (1983) Living with Limitations. Geriatric Nursing 4:107-108.

Hugo, M., R. Goldney, E. Skinner, and M. Katsikitis (1985) Using Screening Instruments In Community Psychiatric Nursing for the Elderly. Australian Journal of Advanced Nursing 2:13-17.

Hutchinson, S. and R.B. Webb (1988) Intergenerational Geriatric Remotivation: Elders Perspectives. Journal of Cross-Cultural Gerontology 3:273-297.

Ide, B. A. (1983) Social Network Syupport among Low-Income Elderly: A Two-Factor Model? Western Journal of Nursing Research 5:233-244.

Janelli, L. M. (1986) The Realities of Body Image. Journal of Gerontological Nursing 12:23-27.

Jenkins, E. H. (1981) Why the Patient Called for Help. Geriatric Nursing 2:37-41.

Johnson, F. L. P. (1979) Response to Territorial Intrusion by Nursing Home Residents. Advances in Nursing Science 1:21-34.

Johnson, F., C. Cloyd, and J. A. Wer (1982) Life Satisfaction of Poor Urban Black Aged. Advances in Nursing Science 4:27-34.

Johnson, F. L., E. Cook, M. J. Foxall, and E. Kelleher (1986) Life Satisfaction of the Elderly American Indian. International Journal of Nursing Studies 23:265-273.

Johnson, J. (1985) Drug Treatment for Sleep Disturbances: Does it Really Work? Journal Gerontological Nursing 11:8-12.

Jones, D. C. and G. M. van Amelsvoort-Jones (1986) Communication Patterns Between Nursing Staff and the Ethnic Elderly in a Long-Term Care Facility. Journal of Advanced Nursing 11:265-272.

Jones, P. L.and A. Millman (1986) A Three-Part System to Combat Pressure Sores. Geriatric Nursing 7:78-82.

Kay, M. A. and C. Tobias (1987) Symptom Care of Widows. Paper presented at the Meeting of the American Anthropological Association Chicago, Il. Nov.18-22.

Kay, M., C. Tobias, B. Ide, and J. Guernsey DeZapien (1988) The Health and Symptom Care of Widows. Journal of Cross-Cultural Gerontology 3:197-208.

Kayser, J. S. and F. A. Minnigerode (1975) Increasing Nursing Students' Interest on Working with Aged Patients. Nursing Research 1:23-26.

Kayser-Jones, J. S. (1979) Care of the Institutionalized Aged in Scotland and the United States: A Comparative Study. Western Journal of Nursing Research 1:190-200.

Kayser-Jones, J. S. (1981a) Old, Alone, and Neglected. Berkeley: University of California Press.

Kayser-Jones, J. S. (1981b) A Comparison of Care in a Scottish and a United States Facility. Geriatric Nursing 2:44-50.

Kayser-Jones, J. S. (1981c) Gerontological Nursing Research Revisited. Journal of Gerontological Nursing 1:217-223.

Keenan, R., A. Redshaw, J. Munson, and W. Mundt (1983) The Benefits of a Drug Holiday. Geriatric Nursing 4:103-104.

Keith, J. (1980) Old Age and Community Creation. In C. L. Fry, Aging in Culture and Society: Comparative Viewpoints and Strategies. Brooklyn, New York: J. F. Bergin Publishers.

Kennedy, S. R. (1985) Sharing, Caring, Living, and Dying (1985) Geriatric Nursing 6:12-17.

Kenworthy, C. (1983) Scared to Death...Almost. Geriatric Nursing 4:358-360.

Kertzer, D. L. and J. Keith (1984) Age and Anthropological Theory. Ithaca, New York: Cornell University Press.

Kim, K. K. (1986) Response Time and Health Care Learning of Elderly Patients. Research in Nursing and Health 9:233-239.

King, F. E., J. Figge, and P. Harmon (1986) The Elderly Coping at Home: A Study of Continuity of Nursing Care. Journal of Advanced Nursing 11:41-46.

Kitson, A.L. (1986) Indicators of Quality in Nursing Care: An Alternative Approach. Journal of Advanced Nursing 11:133-144.

Klein, M., M. Overholser, and H. Rynbergen (1980) Putting Down Roots in Retirement. Geriatric Nursing 1:114-119.

Knowles, L. N. and V. T. Sarver (1985) Attitudes Affect Quality Care. Journal of Gerontological

Nursing 11:35-39.
Kolanowski, L. G. (1981) Hypothermia in the Elderly. Geriatric Nursing 2:362-365.
Kolanowski, A. and L. M. Gunter (1985) What are the Health Practices of Retired Career Women? Journal of Gerontological Nursing 11:22-30.
Kowalsky, E. (1980) Annie. Geriatric Nursing 1:190-192.
LaGioia, K. (1986) Whose Goals Determine Care? Geriatric Nursing 7:190-191.
Lanceley, A. (1985) Use of Controlling Language in the Rehabilitation of the Elderly. Journal of Advanced Nursing 10:125-135.
Lantz, J. M. (1985) In Search of Agents for Self-Care. Journal of Gerontological Nursing 11:10-14.
Lappe, J. M. (1987) Reminiscing: The Life Review Therapy. Journal of Gerontological Nursing 13:12-16.
Laschinger, S. J. (1984) The relationship of Social Support to Health in Elderly People. Western Journal of Nursing Research 6:341-350.
Lashley, M. E. (1987) Predictors of Breast Self-Examination Practice Among Elderly Women. Advances in Nursing Science 9:25-34.
Lederer, A. (1983) Notes on a Nursing Home. Geriatric Nursing 4:224-227.
Leininger, M. (1976) Special Feature: Utah's Dean Madeleine Leininger Says that New Approaches are Needed. Journal of Gerontological Nursing 2:50-51.
Lewis, M. A., S. Cretin, and R. L. Kane (1985) The Natural History of Nursing Home Patients. The Gerontologist 25:382-388.
Lewis, S., R. Messner, and W. A. McDowell (1985) An Unchanging Culture. Journal of Gerontological Nursing 11:20-26.
Lincoln, R. (1984) What Do Nurses Know About Confusion in the Aged? Journal of Gerontological Nursing 10:26-29.
Long, M. L. (1985) Incontinence: Defining the Nursing Role. Journal of Gerontological Nursing 11:30-35.
Lukens, L. (1986) Six Months After Hip Fracture. Geriatric Nursing 7:202-206.
Lund, C. L.and M. L. Sheafor (1985) Is Your Patient About to Fall? Journal of Gerontological Nursing 11:37-41.
Lund, D. A., L. L. Feinhauer, and J. R. Miller (1985) Living Together: Grandparents and Children Tell Their Problems. Journal of Gerontological Nursing 11:29-32.
MacPherson, K. I. (1981) Menopause as Disease: The Social Construction of a Metaphor. Advances in Nursing Science 3:95-113.
MacPherson, K. I. (1985) Osteoporosis and Menopause: A Feminist Analysis of the Social Construction of a Syndrome. Advances in Nursing Science 7:11-22.
Magid, S.and C. R. Hearn (1981) Characteristics of Geriatric Patients as Related to Nursing Needs. International Journal of Nursing Studies 18:97-106.
Martin, C. A. (1984) Aging in Developing Countries. Journal of Gerontological Nursing 10:8-11.
Martinson, I. M. (1982) Does China Really Have the Solution? Journal of Gerontological Nursing 8:263-264.
McBride, M. R. and C. M. Mistretta (1986) Taste Responses from the Chorda Tympani Nerve in Young and Old Fischer Rats. Journal of Gerontology 41:306-314.
McCall, J. (1974) Care of the Elderly in the E.E.C. International Journal of Nursing Studies 11:33-45.
McCracken, A (1987) Emotional Impact of Possession Loss. Journal of Gerontological Nursing 13:14-19.
McKey, P. L. and M. C. Dougherty (1986) The Circumvaginal Musculature: Correlation Between Pressure and Physical Assessment. Nursing Research 35:307-309.
Meguerdichian, D. (1983) Improving Self-Medication in an HRF. Geriatric Nursing 4: 30-34.
Melanson, P. M. and B. L. Downe-Wamboldt (1985) Antecedents of Baccalaureate Student Nurses' Attitudes Toward the Elderly. Journal of Advanced Nursing 10:527-532.
Melanson, P. M. and B. L. Downe-Wamboldt (1987) Identification of Older Adults' Perceptions

of Their Health, Feelings Toward Their Future,and Factors Affecting These Feelings. Journal of Advanced Nursing 12:29-34.
Mentzer, C. A. and J. A. Schorr (1986) Perceived Situational Control and Perceived Duration of Time:Expressions of Life Patterns. Advances in Nursing Science 9:12-20.
Michaelsson, E., A. Norberg, and B. Norberg (1987) A Quality of Life Issue: Feeding Methods for Demented Patients in End Stage of LIfe. Geriatric Nursing 8:69-73.
Miller, A. (1984) Nurse/Patient Dependency: A Review of Different Approaches with Particular Reference to Studies of the Dependency of Elderly Patients. Journal of Advanced Nursing 9:479-486.
Miller, J. (1985) Helping the Aged Manage Bowel Function. Journal of Gerontological Nursing 11:37-41.
Milton, I. and J. MacPhail (1985) Dolls and Toy Animals for Hospitalized Elders: Infantalizing or Comforting? Geriatric Nursing 6:204-206.
Mistretta, C. M. and I. A. Oakley (1986) Quantitative Anatomical Study of Taste Buds in Fungiform Papillae of Young and Old Fischer Rats. Journal of Gerontology 41:315-318.
Mitteness, L. S. (1987) The Management of Urinary Incontinence by Community-Living Elderly. The Gerontologist 27:185-193.
Moore, L. M., C. R. Nielson, and C. M. Mistretta (1982) Sucrose Taste Thresholds: Age-Related Differences. Journal of Gerontology 37:64-69.
Morrisey, S. (1983) Attitudes on Aging in China. Journal of Gerontological Nursing 9:589-593.
Morse, J. M., S. J. Tylko, and H. A. Dixon (1985) The Patient Who Falls...And Falls Again: Defining the Aged at Risk. Journal of Gerontological Nursing 11:15-18.
Morse, J. M., S. J. Tylko, and H. A. Dixon (1987) Characteristics of the Fall Prone Patient. The Gerontologist 27:516-522.
Muhlenkamp, A. F., L. D. Gress, and M. A. Flood (1975) Perception of Life Change Events by the Elderly. Nursing Research 24:109-113.
Nagley, S. J. (1986) Predicting and Preventing Confusion in Your Patients. Journal of Gerontological Nursing 12:27-31.
Neely, E. and M. L. Patrick (1968) Problems of Aged Persons Taking Medications at Home. Nursing Research 17:52-55.
Newman, M. A. (1982) Time as an Index of Expanding Consciousness With Age. Nursing Research 31:290-293.
Newman, M. A. and J. K. Gaudiano (1984) Depression as an Explanation for Decreased Subjective Time in the Elderly. Nursing Research 33:137-139.
Norberg, A. and M. Hirschfeld (1987) Feeding of Severely Demented Patients in Institutions: Interviews with Caregivers in Israel. Journal of Advanced Nursing 12:551-557.
Norton, D., R. McLaren, and A. N. Exton-Smith (1962) An Investigation of Geriatric nursing Problems in Hospital. National Corporation for the Care of Old People, London.
Osborne, O. H. (1977) Aging and the Black Diaspora: The African, Caribbean, and Afro-American Experience. In M. Leininger, ed., Proceedings: Transcultural Nursing Care of the Elderly. Second National Transcultural Nursing Conference.
Pajk, M., G. A. Craven, J. Cameron-Berry, and T. Shepps (1986) Investigating the Problem of Pressure Sores. Journal of Gerontological Nursing 12:11-16.
Palmateer, L. M. and J. R. McCartney (1985) Do Nurses Know When Patients Have Cognitive Deficits? Journal of Gerontological Nursing 11:6-16.
Parent, C. J. and A. L. Whall (1984) Are Physical Activity, Self Esteem, and Depression Related? Journal of Gerontological Nursing 10:8-11.
Pearson, B. D. (1987) Pain Control: An Experiment with Imagery. Geriatric Nursing 8:28-30.
Pease, R. A. (1985) Praise Elders to Help Them Learn. Journal of Gerontological Nursing 11:16-20.
Peay, D. M. (1977) Some Cultural Values of Mormons and Their Implications for Health Care of the Elderly. In M.Leininger, ed. Proceedings: Transcultural Nursing Care of the Elderly.

Second National Transcultural Nursing Conference.
Penner, L. A., K. L. Ludenia, and G. Mead (1984) Staff Attitudes: Image or Reality? Journal of Gerontological Nursing 10:110-117.
Pensiero, M. and M. Adams (1987) Dress and Self Esteem. Journal of Gerontological Nursing 13:10-17.
Petrou, M. F. and J. V. Obenchain (1987) Reducing Incidents of Illness Post Transfer. Geriatric Nursing 8:264-266.
Phillips, L. R. (1983) Abuse and Neglect of the Frail Elderly at Home: An exploration of Theoretical Relationships. Journal of Advanced Nursing 8:379-392.
Phillips, L. R. and V. F. Rempusheski (1986) Caring for the Frail Elderly at Home: Toward the Theoretical Explanation of the Dynamics of Poor Quality Family Caregiving. Advances in Nursing Science 8:62-84.
Podskalny, L. K. and D. Woods (1983) British Geriatric Health Care. Journal of Gerontological Nursing 9:604-613.
Pohl, J. M. and S. S. Fuller (1980) Perceived Choice, Social Interaction, and Dimensions of Morale of Residents in a Home for the Aged. Research in Nursing and Health 3:147-157.
Pollman, J. W., J. Morris, and P. N. Rose (1978) Is Fiber the Answer to Constipation in the Elderly? A Review of the Literature. International Journal of Nursing Studies 15:101-104.
Porter, S. H. (1984) Collaborating with the World Health Organization Journal of Gerontological Nursing 10:30-34.
Powers, B. A. (1988a) Social Networks, Social Support, and Elderly Institutionalized People. Advances in Nursing Science 10:40-58.
Powers, B. A. (1988b) Self-Perceived Health of Elderly Institutionalized People. Journal of Cross-Cultural Gerontology 3:299-321.
Preston, D. and J. Grimes (1987) A Study of Differences in Social Support. Journal of Gerontological Nursing 13:36-40.
Rader, J. (1987) A Comprehensive Staff Approach to Problem Wandering. The Gerontologist 27:756-760.
Rader, J., J. Doan, and M. Schwab (1985) How to Decrease Wandering: A Form of Agenda Behavior. Geriatric Nursing 6:196-199.
Ragucci, A. T. (1977) The Urban Context of Health and Illness Beliefs and Practices of Elderly Women in an Italian-American Enclave. In M.Leininger, ed. Proceedings: Transcultural Nursing Care of the Elderly. Second National Transcultural Nursing Conference.
Reed, P. G. (1983) Implications of the Life-Span Developmental Framework for Well-Being in Adulthood and Aging. Advances in Nursing Science 6:18-25.
Reed, P. G. (1986) Developmental Resources and Depression in the Elderly. Nursing Research 35:368-374.
Remondet, J. H. and R. O. Hansson (1987) Assessing a Widow's-Grief: A Short Index. Journal of Gerontological Nursing 13:30-34.
Rempusheski, V. F. (1988) Caring for Self and Others: Second Generation Polish American Elders in an Ethnic Club. Journal of Cross-Cultural Gerontology 3:223-271.
Ricci, M. (1983) All-Out Care for an Alzheimer Patient. Geriatric Nursing 4:369-371.
Richter, J. M. (1984) Crisis of Mate Loss in the Elderly. Advances in Nursing Science 6:45-54.
Richter, J. M. (1987) Support: A Resource During Crisis of Mate Loss. Journal of Gerontological Nursing 13:18-22.
Rigdon, I. S., B. C. Clayton, and M. Dimond (1987) Toward a Theory of Helpfulness for the Elderly Bereaved: An Invitation to a New Life. Advances in Nursing Science 9:32-43.
Robb, S. S. (1979) Attitudes and Intentions of Baccalaureate Nursing Students Toward the Elderly. Nursing Research 28:43-50.
Robb, S. S. (1983) The Challenge of Research. Journal of Gerontological Nursing 9:336-343.
Robb, S. S. (1985) Urinary Incontinence Verification in Elderly Men. Nursing Research 34:278-282.

Robb, S. S., M. Boyd, and C. L. Pristash (1980) A Wine Bottle, Plant, and Puppy: Catalysts for Social Behavior. Journal of Gerontological Nursing 6:721-28.
Robb, S. S. and C. E. Stegman (1983) Companion Animals and Elderly People: A Challenge for Evaluators of Social Support. The Gerontologist 23:277-282.
Robinson, L. (1981) Gerontological Nursing Research. In I.Burnside ed Nursing and the Aged. (Second edition). New York: McGraw Hill.
Robinson, S. B. and P. L. Demuth (1985) Diagnostic Studies for the Aged: What are the Dangers? Journal of Gerontological Nursing 11:6-12.
Roosa, W. M. (1982) Territory and Privacy: Residents' Views. Geriatric Nursing 3:241-243.
Ryden, M. B. (1984) Morale and Perceived Control in Institutionalized Elderly. Nursing Research 33:130-136.
Ryden, M. B. (1985) Environmental Support for Autonomy in the Institutionalized Elderly. Research in Nursing and Health 8:363-371.
Safier, G. (1976) Oral Life History with the Elderly. Journal of Gerontological Nursing 2:16-22.
Salisbury, S. and P. Goehner (1983) Separation of the Confused or Integration with the Lucid? Geriatric Nursing 4:231-233.
Sands, D. and E. Holman (1985) Does Knowledge Enhance Patient Compliance? Journal of Gerontological Nursing 11:23-29.
Saunders, V. (1984) Profiles of Elderly Armenians. Journal of Gerontological Nursing 10:26-29.
Schorr, J. A. (1983) Manifestations of Consciousness and the Developmental Phenomenon of Death. Advances in Nursing Science 6:26-35.
Schwab, M., J. Rader, and J. Doan (1985) Relieving Fear and Anxiety in Dementia. Journal of Gerontological Nursing 11:8-15.
Schwartz, D. (1980) Hamlet Dweller – City Dweller. Geriatric Nursing 1:128-132.
Schwartz, D. (1982) Catastrophic Illness: How It Feels. Geriatric Nursing 3:303-306.
Seabrooks, P. A., R. Kahn, and G. Gero (1987) Cross-Cultural Observations. Journal of Gerontological Nursing 13:18-22.
Secrest, V. (1984) A Tribute to Lady Joanna. Geriatric Nursing 5:19-21.
Sexton, D. L. (1984) The Supporting Cast: Wives of COPD Patients. Journal of Gerontological Nursing 10:82-85.
Shannon, M. (1976) A Measure of the Fulfillment of Health Needs. International Journal of Nursing 7:353-377.
Shelly, S. I., R. M., and C. D. S. Gambrill (1987) Aggressiveness of Nursing Care for Older Patients and Those with Do-Not-Resuscitate Orders. Nursing Research 36:157-162.
Shimamoto, Y. (1977) Health Care to the Elderly Filipino in Hawaii. In M. Leininger, ed. Proceedings: Transcultural Nursing Care of the Elderly. Second National Transcultural Nursing Conference.
Shimamoto, Y (1984) Aging in Palau. Journal of Gerontological Nursing 10:13-16.
Shimamoto, Y. and C. L. Rose (1987) Identifying Interest in Gerontology Journal of Gerontological Nursing 13:8-13.
Shomaker, D. (1979) Dialectics of Nursing Homes and Aging. Journal of Gerontological Nursing 5:45-48.
Shomaker, D. (1987) Problematic Behavior and the Alzheimer Patient: Retrospection as a Method of Understanding and Counseling. The Gerontologist 27:370-375.
Simon, J. M. (1987) Health Care of the Elderly in Appalachia. Journal of Gerontological Nursing 13:32-35.
Simons, J. (1985) Does Incontinence Affect Your Client's Self-Concept? Journal of Gerontological Nursing 11:37-42.
Slimmer, L. W., M. Lopez, J. LeSage, and J. R. Ellor (1987) Perceptions of Learned Helplessness. Journal of Gerontological Nursing 13:33-37.
Smith, C. E., S. Buck, E. Colligan, and P. Kerndt. (1980) Differences in Importance Ratings of Self-Care Geriatric Patients and the Nurses Who Care For Them. International Journal of

Nursing Studies 17:145-153.

Smith, D. L. and A. E. Molzahn-Scott (1986) A Comparison of Nursing Care Requirements of Patients in Long)Term Geriatric and Acute Care Nursing Units. Journal of Advanced Nursing 11:315-321.

Smith, S., V. Jepson, and E. Perloff (1982) Attitudes of Nursing Care Providers Toward Elderly Patients. Nursing and Health Care 3:93-98.

Snape, J. (1986) Nurses' Attitudes to Care of the Elderly. Journal of Advanced Nursing 11:569-572.

Spotts, S. J. (1981) Love: Crisis and Consolation for the Middlescent Woman. Advances in Nursing Science 3:87-94.

Stevenson, J. and P. Gray (1981) Rehabilitation for Long-Term Residents. Geriatric Nursing 2:127-131.

Strumpf, N. E. (1986) Studying the Language of Time. Journal of Gerontological Nursing 12:22-26.

Strumpf, N. E. (1987) Probing the Temporal World of the Elderly. International Journal of Nursing Studies 24:201-14.

Sullivan, J. A. and F. Armignacco (1979) Effectiveness of a Comprehensive Health Program for the Well-Elderly by Community Health Nurses. Nursing Research 28:70-75.

Sullivan, R. T. (1977) Some Values, Beliefs, and Practices of Elderly Women in The United States. In M. Leininger, ed. Proceedings: Transcultural Nursing Care of the Elderly. Second National Transcultural Nursing Conference.

Taft, L. B. (1985) Self-Esteem in Later-Life: A Nursing Perspective. Advances in Nursing Science 8:77-84.

Taylor, K. H. and T. L. Harned (1978) Attitudes Toward Old People: A Study of Nurses Who care for the Elderly. Journal of Gerontological Nursing 4:43-47.

Taylor, K. and J. Henderson (1986) Effects of Biofeedback and Urinary Stress Incontinence in Older Women. Journal of Gerontological Nursing 12:25-30.

Thornbury, J. M. and C. M. Mistretta (1981) Tactile Sensitivity as a Function of Age. Journal of Gerontology 36:34-39.

Thornbury, J. M. and A. Martin (1983) Do Nurses Make a Difference? Journal of Gerontological Nursing 9:440-445.

Tien-Hyatt, J. L. (1986-87) Self-Perceptions of Aging Across Cultures: Myth or Reality? International Journal of Aging and Human Development 24:129-148.

Timan, B., D. Goldfarb, and B. Curtis (1982) Home Safe. Geriatric Nursing 3:399-401.

Tobiason, S. J., F. Knudsen, and J. C. Stengel (1979) Positive Attitudes Toward Aging: The Aged Teach the Young. Journal of Gerontological Nursing 5:18-23.

Tripp-Reimer, T. (1980) Cultural Perspectives on Aging. In Holistic Assessment of the Healthy Aged. M. M. Schrock, ed. New York: John Wiley and Sons.

Tripp-Reimer, T. and M. C. Dougherty (1985) Cross-Cultural Nursing Research. In Annual Review of Nursing Research. H. H. Werley and J. J. Fitzpatrick, eds. New York: Springer.

Tripp-Reimer, T., B. Sorofman, G. Lauer, M. Martin, and L. Afifi (1988) To Be Different From the World: Patterns of Elder Care Among Iowa Old Order Amish. Journal of Cross-Cultural Gerontology 3:185-195.

Turkoski, B. B. (1985) Growing Old in China. Journal of Gerontological Nursing 11:32-34.

Valanis, B., R. C. Yeaworth, and M. R. Mullis (1987) Alcohol Use Among Bereaved and Nonbereaved Older Persons. Journal of Gerontological Nursing 13:26-32.

Venglarik, J. M. and M. Adams (1985) Which Client is at Higher Risk? Journal of Gerontological Nursing 11:28-35.

Verhonik, P. J. (1961) Decubitus Ulcer Observations Measured Objectively. Nursing Research 10: 211-214.

Wade, B. and P. Snaith (1981) The Assessment of Patients' Need for Nursing Caree on Geriatric Wards. International Journal of Nursing Studies 18: 261-271.

Wade, B. and A. Bowling (1986) Appropriate Use of Drugs by Elderly People. Journal of Advanced Nursing 11: 47-55.

Wagnild, G. and R. W. Manning (1985) Convey Respect During Bathing Procedures: Patient Well-Being Depends on It. Journal of Gerontological Nursing 11:6-10.

Waters, K. R. (1987a) Discharge Planning: An Exploratory Study of the Process of Discharge Planning on Geriatric Wards. Journal of Advancved Nursing 12: 71-83.

Waters, K. R. (1987b) Outcomes of Discharge from Hospitals for Elderly People. Journal of Advances Nursing 12: 347-355.

Wells, T. J. (1984) Social and Psychological Imnplications of Incontinence. In Urology in the Elderly. J. C. Brocklehurst, ed. Edinburgh: Churchhill Livingstone.

Wells, T. J., C. A. Brink, and A. C. Diokno (1987) Urinary Incontinence in Elderly Women: Clinical Findings. Journal of the American Geriatrics Society 35: 940-946.

White, H. E., N. E. Thurston, K. A. Blackmore, and S. E. Green (1987) Body Temperature in Elderly Surgical Patients. Research in Nursing and Health 10 317-321.

White, P. H. (1983) Supportive Counsling in Action. Geriatric Nursing 4: 176-177.

Whitney, J. D., B. J. Fellows and E. Larson (1984) Do Mattresses Make a Difference? Journal of Gerontological Nursing 10: 20-25.

Williams, A. (1972) A Study of Factors Contributing to Skin Breakdown. Nursing Research 21: 238-243.

Williams, M. A., J. R. Holloway, M. C. Winn, and M. O. Wolanin (1979) Nursing Activities and Acute Confusional States in Elderly Hip-Fractured Patients. Nursing Research 28: 25-35.

Williams, M. A., E. B. Campbell, W. J. Raynor, and S. M. Mlynarczyk (1985) Predictors of Acute Confusional States in Hospitalized Elderly Patients. Research in Nursing and Health 8: 31-40.

Wilson, H. S. and J. Heinert (1977) Los Viejitos: The Old Ones. Journal of Gerontological Nursing 3: 19-25.

Wilson, J. (1981) The Plight of the Elderly Alcoholic. Geriatric Nursing 2: 114-118.

Wiltzius, F., S. R. Gambert and E. H. Duthie (1981) Importance of Resident Placement Within a Skilled Nursing Facility. Journal of the American Geriatrics Society 29: 418-421.

Wirtz, B. J. (1987) Effects of Air and Water Matresses on Thermo-Regulation. Journal of Gerontological Nursing 13: 13-17.

Wolanin, M. O. (1981) Nursing Therarpy, Drug Therapy or Both? Geriatric Nursing 2: 408-410.

Wolanin, M. O. (1983a) Confusion: Scope of the Problem and Its Diagnosis. Geriatric Nursing 4: 227-230.

Wolanin, M. O. (1983b) Clinical Geriatric Nursing Research. In Annual Review of Nursing Research. H. H. Werley and J. J. Fitzpatrick, eds. New York: Springer.

Yarmesch, M. and M. Sheafor (1984) The Decision to Restrain. Geriatric Nursing 5: 242-244.

Yu, L. C. (1987) Incontinence Stress Indes: Measuring Psychological Impact. Journal of Gerontological Nursing 13: 18-25.

Yu, L. C. and D. L. Kaltreider (1987) Stressed Nurses: Dealing with Incontinent Patients. Journal of Gerontological Nursing 13: 27-30.

# SECTION TWO

# CULTURAL ISSUES

ROBERT L. RUBINSTEIN

## 4. NATURE, CULTURE, GENDER, AGE: A CRITICAL REVIEW

INTRODUCTION

Gender forms one of the most important variables in human aging because it is one of the most important aspects of human experience. This is a review essay on gender and age, but it will depart from the usual topical treatment by discussing these two constructs as concepts, as methodologic tools, and as orienting assumptions about the world as lived and understood.

This is not to say that a topical treatment of gender and age from an anthropological perspective would not be a useful enterprise. A topical review would need to reflect on the vast amount of recent literature on gender that touches on the elderly. In the West, there is a growing literature on older men (Farrell and Rosenberg 1981; Levinson 1978; Lewis and Salt 1986; McMorrow 1974 Pesman 1984; Rubinstein 1986a) and on women (Alexander 1986; Bell 1986; Block, Davidson and Grambs 1981; Finch and Groves 1983; Formanek and Gurian 1987; Gee and Kimball 1987; Gibson, 1985; Harris 1975; Haug, Ford, and Sheafor 1985; Hawley 1985; Henig 1985; Lopata 1979; Melamed 1983; Nett 1982; Nudel 1978; Szinovacz 1982; Silverman 1981; Simon 1987; Walker 1985) and in which gender forms a key issue. There is also a growing literature on non-Western societies on gender and gender ideology. At the very least it would need to discuss questions of comparative longevity and social status and power through the life span.

But there is more than this. Simply put, there are few studies of later life in which gender – of a sample or of an individual – is not a significant datum. Yet there are many ways in which gender and age have been treated by investigators. Each perspective concerning gender represents tacit, operating assumptions about gender and the "givenness" of gender. Rosaldo (1987:300) notes how given presumptions about gender are represented in many orienting dichotomies in the social and behavioral sciences: the domestic and public domains, the psychological and the social, nature and nurture, morality and competition. She notes, "All depend on more or less specified assumptions concerning sex....[and] none helps us understand the place of gender in the ways we think not just of sex but of such diverse things as youth and age..."

In examining tacit operating assumptions about gender and age here, I will explore some issues that have been neglected, for the most part, in cultural studies of later life, but that seem to me to be critical in our understanding of the cultural meaning of aging. The issues raised here have to do with the relationship of aging, gender, "nature" and biology. The question I wish to pose is this: Why does aging seem to us to be more biological ("natural") than cultural? This view,

I feel, is widely represented in both academic and popular culture in the West. The point is this: if "biology" is our significant folk construct – as it has been shown to be – and if our perception of aging is cultural, isn't our perception of aging as fundamentally biological a cultural perception? Related to this is the possibility that the range of cultural perceptions of later life and the meanings assigned to old people, their behavior and capacities can be conceptualized, alternatively, as "natural" or as "cultural." Below, in the first part of this essay, I will discuss various permutations of the culture and nature theme in studies of gender and age as these intertwine in scholarly work and in our own folk models. In the second part, I will describe three ways in which cultural thinking has, or has not, been a part of research on aging.

## NATURE AND CULTURE

In this essay I am thus concerned with the observation made by Collier and Yanagisako (akin to that of Rosaldo) that our folk system posits that behavioral differences not explained by culture (or "society") must be due to nature (or "biology") and vice versa.

The stepping off point here is the recent discussion about the cultural nature of gender. In posing the question, Is man to culture as women is to nature, and by suggesting that cultures by and large follow this distinction, Ortner (1974) posited a general model for symbolic classification of types of humans and related overarching sociocultural domains. The universality of female subordination, she hypothesized, is based on the identification of women with "nature....something that every culture devalues, something that every culture defines as being of a lower order of existence than itself" (1974:72). In contrast, men are identified with "culture," "the process of generating and sustaining systems of meaningful forms" (1974:72). Why should women be viewed in this way? Because of "the body and the natural procreative functions specific to women alone" (1974:72) at three analytical levels: woman's bodies and their functions that engender a perception as closer to nature; the lower and fewer social roles accorded them; and differences in psychic structure ensuing from nature-proximity and lesser social roles.

Whatever the relationship between gender, nature, and culture, it is clear that, in our own folk model at least, these intertwine with age. At another level, in an American folk model of the life course (of aging) an equation is made of both youth and old age with "nature" and mid-life with "culture." From this perspective, the aged are perceived as "natural" in that they are conceptualized as prisoners of time, closer to death, less resistant to disease, subjected to emotional swings and radical emotions such as depression, made acultural through isolation, stripped through loss, all in all, more greatly affected by biology and other natural phenomena. In contrast, in mid-life, the "boundary" between the cultural person and nature is thicker and clearer. The person is defined as more in control, more concretely in a cultural role, and less subject to the depredations of "nature."

It is clear that Ortner's gender model, as a scientific model, is not correct although it may in fact be *our* folk model of gender (MacCormack 1980). Yet whatever its correctness however, its proposal led in two significant directions.

First, it has brought about a good deal of discussion and debate, reformulation, and continuing productive research. For example, MacCormack (1980) has decried the ethnocentricity and reductionism of the distinction between nature and culture as applied to gender. In responding to Ortner's assertion that everywhere women are considered in some degree inferior to men, MacCormick notes, "But [Ortner] does not say by whom they are considered to be so. By men? By women? By how many?" (1980:17). She believes that many of her women field informants would disagree with Ortner (cf. Goodale 1980). M. Strathern (1980:177) notes too "There is no such thing as nature or culture. Each is a highly relativized concept whose ultimate signification must be derived from its place within a specific metaphysics." For Westerners, as Schneider and others have noted, nature and culture are key elements of our own system of folk classification.

The second development from this approach is a greater and more painstaking attention to ethnographic reality that has led to a liberating emphasis on the perception of gender as truly a socioculturally-constructed symbolic subsystem, existing in historic process, rather than as a constricting biological given. As Ortner and Whitehead note, "Natural features of gender, and natural processes of sex and reproduction, furnish only a suggestive and ambiguous backdrop to the cultural organization of gender and sexuality. What gender is, what male and female are, what sorts of relations do or should obtain between them – all of these notions do not simply reflect biological 'givens,' but are largely the product of social and cultural processes" (1981:1). Noting, too, that male and female are "predominantly cultural constructions" rather that "predominantly natural objects", they suggest that an adequate understanding of gender begins through "asking what male and female, sex and reproduction, *mean* in given social and cultural contexts, rather than assuming we know what they mean in the first place" (1981:1), because each of these, "male," "female," "sex," and "reproduction" is a cultural symbol, linked to and only truly graspable through its link to other cultural symbols.

Thus, a crucial part of our own understanding of gender has had to do with the realization of the limits of our own analytical categories (such as "male," "female," "nature" and "culture") particularly as these are based on our own cultural categorizations. As some of the notions described above clearly indicate, biological thinking has greatly muddled understanding of gender as a sociocultural thing. Social science understanding is closely related to the degree to which our own analytical constructs are free of both conscious and unconscious biases. Significant in this respect are two of our own cultural assumptions.

The first is our own scientific reliance on Western folk notions of biology as an explanation for sociocultural phenomena and related to this the genderization of science (Bleier 1984; Fausto-Sterling 1985; Sahlins 1976).

The second is the cultural pervasiveness of our own folk notions of biology as these are a part of the distinction between nature and culture, which is itself a western folk model (Strathern 1980). This has been particularly evident in traditional anthropological studies of kinship. Schneider, in his essay, "What Is Kinship All About" (1972) and in many other works, has noted that "kinship" as it has been used in anthropology *"does not correspond to any cultural category known to man"* (1972: 50; italics in original). This is because of the preoccupation of traditional kinship studies with genealogy and thus with the folk biological model of relationships from Western culture as they have been incorporated into the analytical constructs of our social science. He argues for the importance of analyzing specific cultural symbols (such as nature and culture in American kinship) in order to understand the cultural meanings of human relationships (cf. Luborsky 1987, concerning retirement). We can not presume that our own terms supply us with an exact translation of others' concepts.

CULTURE AND AGE

We are now able to turn to a significant related question. If we pose the question, "Is culture to nature as gender is to age?" we must answer, "Yes, but it shouldn't be." While we have increasingly come to recognize gender-based elements such as "male," "female," and "reproduction" as cultural constructs, we have, by and large, failed to do so with basic aging constructs, particularly "chronology."

Despite the basis of gender differences in the "suggestive backdrop" of the biology of sex, viewing "sexual meanings" as sociocultural and symbolic constructs now seems relatively an easy task for social scientists. At the very least, because of feminist scholarship in anthropology, we are aware of the potential for male bias in fieldwork and analysis, of the wide range of gender roles, meanings and symbols in the world's cultures, and of the types of relationships among gender symbols, gender-based roles, practices, behaviors, ideological conceptions, and historic and power circumstances. This awareness has followed some clear changes in gender-based social roles in contemporary American life and increased public discussion and consideration of these.

Yet, if my reading of things is correct, it seems to me more difficult to view old age (in comparison with gender) as socioculturally freed of its biological context. Myerhoff (1984:307) has written, "As humans, we all are born, mate and die; but we manipulate these constancies endlessly, so that biology with all its imperatives and universals can often be only faintly distinguished beneath the templates of symbolic and ritual forms that overlap it." Yet *our* view of the life course, our "template of symbolic form" or, at least, the glass through which we see, is our folk biology and the object of our view, as both scientists and citizens, is changed hue in that way.

There are several circumstances at work here. Certainly, our cultural biologization of old age is part and parcel of our popular stereotypes of aging. These common images, some positive but mostly negative, portray older persons as prisoners of biology in a variety of ways. Elders may be viewed as infant-like or dependent ("in a second childhood"; Arluke and Levin 1985); as fully preoccupied with health and mortality; as subject to "declines" such as memory loss and cognitive difficulties); resting on a rocker after the labors of life; as incompetent or decrepit; and as otherwise subject to the whims of "nature." Sankar (1984) notes that according to the Harris poll on aging, Americans in general believe that the primary cause of later life disabilities is old age itself, which is culturally viewed "as a kind of disease" (1984:251), an evaluation shared by a medical establishment characterized by "physician antipathy to the elderly" (1984:257). In anthropology, like Western culture at large, there is often a tendency to separate things that are perceived as negative and to attribute them to innate human nature, as Siskind (1978) notes.

But there is more than popular cultural imagery. The very term "old age" is thick with cultural assumptions about chronology, the linear nature of time, decline, as well as "natural" change, and the life course with the assumption of analogies to nature (seasons, gardens, fires), loss, the diminution of life space, and the infringement upon the person by the forces of nature. Ostor (1984:281) has written that "age renders itself to social scientists in the West with such force and seeming objectivity that we assign it an almost universal linearity, seen in terms of stages of development, and an unvarying reality across cultures." Ostor suggests that a solution is focusing on problems of cultural meaning of aging and development as well as time which may be alternatively viewed as "cyclical, plural, reversible, nonlinear, and nonmeasureable" (1984:282).

There is abundant evidence that time, temporality, the passage and marking of time are cultural constructs. Yet for us, old age and linear chronology implicate one another and are a part of the phenomenological field and our natural understandings of our everyday view of the aged. Any attempt to culturalize old age must first confront its underlying basis in our cultural construction of "chronology" (see below).

So we must keep in mind that both gender and age have the same bases in biology and provide the same "amorphous" background material for the operation of powerful cultural symbols such as "nature," "culture" and others. It hardly seems possible that old age is more innately "biological" than gender.

Viewing gender as a cultural construct, freed from its scientific and popular enmirement in cultural biology, is a task perhaps made easier by the seeming simplicity of the binary pair: male-female. From a biological perspective one *is* one or the other (although this is not the case culturally; cf.

Ortner and Whitehead 1981). In addition, there is an added conceptual separation: one may speak of sex for biology and gender for sociocultural meaning (Shapiro 1981). Such an orienting duality is at present absent from the study of the life course. There is no such conceptual language in the study of age. For example a distinction between "biological age" and "social age" does not quite compel conceptual clarity in that the term "age" is itself fraught with thick cultural meanings. Fortes (1984) emphasizes the failure of the correlation of life course stages and chronology and he notes too that anthropologists often project the interpretation of chronological age as significant or as an ordering frame on materials from field data. And there is no uniform conception, biologically or culturally of when old age begins although one is, biologically at least, gendered from birth. One possible analogy to sex/gender is chronology/aging.

## THE NATURE OF GENDER

On first scrutiny, gender is seen to form a simple dichotomous category that crossects age or lifespan which itself is said to feature a universally-occurring three stage minimum: childhood, adulthood and old age. There is mounting evidence that as far as specific cultural meanings are concerned, age and gender – our western folk cultural and scientific categories – merge or meld. In one sense, age can be thought of as being overridden by gender, as Counts and Counts (1985:7) note: "men and women experience old age differently in so many societies." Perhaps then there are no constant effects of "aging" on individuals in society at all that are not strongly modified by how they are defined as culturally gendered persons. More radically, we may even cut ourselves adrift from such correlational thinking and suggest that when we pose the question "What is old age all about?" the answer is that it is quite often about gender and culturally specific gender symbolism.

*Where is the culture in aging research?*

Thus by and large, when we ask what gender is we respond "culturally" and when we ask what aging is we respond "naturally." In the remainder of this essay, I will address the question, "Where is the culture in aging research?"

The answer is, Not in as many places as it should be. If we examine, with a concern for culture, general research on aging, we are able to discern three overarching perspectives. I have labeled these here the acultural, the culturalized, and the cultural.

Acultural analysis, typical of mainstream social gerontology, fails to consider culture, meaning, and cultural process. By hook or by crook, it represents a verbal version of stimulus response psychology applied to the study of old age. Gender becomes one of a number of background variables to be examined, whose meaning is operationally fixed or assumed, or emergent only

NATURE, CULTURE, GENDER, AGE: A CRITICAL REVIEW   115

from statistical analysis, as if it is wholly devoid of social meaning. The culturalized approach, not uncommon in many disciplines and quite common in anthropology, follows the logic of our own folk system in its view of nature and culture, taking the "meaning" of gender and age (our meaning of course) as a given. The constructed nature and meanings attached to gender and age are rarely examined. The variety of situated cultural meanings germane to age, the cultural definition of the relationship among ages, the existences of intermediate statuses, change over the life-span, and the problems inherent in translating others' age notions into "our" age notions need to be more carefully addressed.

Only recently however are we beginning to think through the effects of our own folk beliefs about the nature of things, and to work to develop a truly cultural perspective that can be extended to the study of "old age." From this approach we begin to view as primary the arbitrary nature of cultural symbols, to the point of analytically assuming their separateness from the basic cultural categories with which they are habitually associated in both Western folk and in scientific usage.

*The Acultural Approach*

This perspective is closest to that of mainstream social gerontology. It suggests that to whatever extent the lives of older men and women are different in North American and other societies is due to the later life residuum of social roles and practices and innate biological differences. It employs the language of measurement. It assumes differences and similarities can be "measured" by treating age and gender as discrete variables and as uniformly equal, comparable chunks of social identity (cf. Beeson 1975).

There appear to be two ways this might occur: gender comparison across some dimensions and the blurring of gender distinctions. For example in a recent paper Kendig *et al.* (1987) compare older men and women in an examination of "confidantes and family structure in old age" and note that the gender of the respondent and the gender of close family members emerged as important influences on "confiding patterns" in a study of 1,050 persons age 60 and older. For example, we learn that "except for the oldest age group, husbands were generally more likely than wives to confide in their spouses" (1987:S38). Gender, they note, influences the number of confidants in the oldest age group, with women having more confidants than men. They note, too, that their findings "are generally consistent with literature suggesting that men are more dependent than women on intimacy within marriage and that women have social skills that facilitate confiding in other relationships, but they also indicate that these effects are by no means universal" (1987:S38).

This paper is typical of much social gerontology literature in approach and narration of findings. Yet these are somehow unsatisfactory for an anthropologist. The "findings" appear as decontextualized, as stripped from setting and meaning. These are findings that are divorced from the reality of gender as a meaningful category or as a lived experience. We learn basically that older husbands confide in their wives. An anthropologist would wish to know more about the cultural and personal meaning of these gender interactions.

Another striking example of acultural analysis across some dimension (in my mind) is contained in the decades-long debate over who does worse, older widowers or widows (Arens 1982; Atchley 1975; Berardo 1968, 1970; Gallagher et al. 1983; Heyman and Gianturco 1973; Helsing, Szklo and Comstock 1981; Hyman 1983; Morgan 1976; Young, Benjamin and Wallace 1963). This set of papers demonstrates an increasing technical sophistication in conceptualizing and measuring post-loss morale and symptomatology. Yet many anthropologists would wish to raise issues about the cultural contexts in which emotions are experienced and expressed. Indeed, some might note that when we compare male and female in such a manner, we are actually comparing entities that are significantly different on the basis of experienced affect and operating emotional systems. And, if the outcome variables such as distress or morale are in fact not significantly different for bereaved older widowers and widows, anthropologists might also question whether the processes of reorganization, ways of grappling with grief, access to new experiences, gender and ethnic-based expressive patterns, and modes of incorporating the past that lead to the equivalent outcomes are themselves the same. And if they aren't, shouldn't they be studied as well?

The second type of acultural approach refers to the blurring of gender and age. This is also of interest because of how such categoric blending separates the social status of all elders *as* a social category, regardless of gender. Here, gender differences are seen as strongly blurred by age primarily because of the social emphasis on the linearity of age in Western societies, our own confusion of gender with sex, and the operation of the folk model – described above – that links later life with nature. In this view people aren't first men and women. Rather, they are first old (and closer to "nature") and gender (defined in a culturally essentialist manner as reproductive ability or sexual tone) is of secondary significance. A question is this: If there are so many culturally-based gender differences between men and women in later life (in social stressors relating to longevity, emotionality, experience of illness, modes of relating to other people, income, past experiences as culturally structured in the life course, to name a few), how can these be ignored? Yet we frequently talk of the elderly as a uniform or discrete group.

*The Culturalized Approach*

The second perspective on aging (and gender) can be described as culturalized. This term has been chosen to emphasize an overall orientation to the significance of cultural meaning and differences and yet to recognize a continuing difficulty in maintaining a stance of cultural relativity.

While this approach is exemplified in the work of very many cross-cultural researchers, I wish to discuss at some length and critique here the work of Gutmann. In a recent book (1987), he suggests that old age derives its essential human meaning from its species role as a post-parental phenomenon, developing ultimately from the "natural" parental functions. In order to successfully raise children and confront their initial helplessness, the parents themselves may require nurturance typically from the elders of their communities. The role of elders in history has been to foster an environment for effective parenting and so elders help maintain the evolutionary advantage of the human brain and the propensity to new learning, Gutmann finds. In their roles, human parents are backed up by grandparents who act as caretakers of the human heritage.

According to Gutmann, as men age, they are able to direct the unconscious content of fantasy into conventional forms of communication and expression in the outer world. In projective tests administered cross-culturally, he found a diminution of the imagery of aggression by men with age. He supports these findings with ethnographic data from age-graded societies of Africa (in which young men are perceived as killers and older men as peacemakers) and from American Indian and Asian materials. He discusses, briefly, the transition for men from "warrior to 'woman'" (a change that may be particularly a characteristic of macho cultures). He notes that the general structure is that older men become more like women and increasingly spend time with women. Becoming domesticated, men also retreat into the domestic sphere and space, described as more internalized and female. Clinically, this is manifested in such elements as "male pacification," "inert postures," "self-nurturing," and the (psychoanalytic) view that stresses "the aging male's withdrawal from active engagement with the world" (1987:97) leading to the expression of sensual, maternal and oral concerns.

In contrast, according to Gutmann, the aging woman experiences a degree of inner freedom. He finds that both the psychological and ethnographic evidence he has reviewed support the notion that women do not copy but rather reverse the structure of how men age (1987:133). Women begin their lives with the psychological characteristic of Passive Mastery prominent, dependent and even deferent to the husband, but end up with a disposition to Active Mastery as a psychological modality. These changes hint of supplementation rather than diminution. Such changes are not made independent of those in men, Gutmann asserts, but the two are bound together in a dynamic process in later life. Some findings he cites: middle age American women become more dominant in the family; the existence of stereotypes of powerful women in popular American culture; the role of work as a significant factor in the lives of older women; the role of older women as ritual leaders in traditional societies (cf. Brown 1985); the acquisition of a degree of political and social freedom, perhaps for the first time in life; the role of mother-in-law as a locus of power; and the existence of a family alliance between the matriarch and the eldest son. In contrast to these generally positive views and socially sanctioned roles for women in later life are different images of women with "active mastery" including the common view of the older women as witch, as hag or as evil spirit. In many ways, this is an interesting twist on the American folk model of aging, with the natural tendencies of the genders, reined in during middle life, "naturally" emergent in later life, but in gender-specific ways.

He relates these developmental differences in life span gender proclivities to issues of parenting. Parents must sacrifice to raise children. Parents will, naturally and developmentally, fundamentally alter and restrict their own needs and behaviors and reconstitute their own psychological profiles in order to serve the needs of children. According to Gutmann, such needs are encompassed by two types of security, physical and emotional, the provision of which was evolutionarily secured by the sorting of capacities by sex and their attendant universality in human cultures, so that important tasks are accomplished. The universality of the division of parental tasks and the development in these tasks of the most salient forms of gender role patterning occur as part of and in reaction to the "chronic emergency" of parenthood, he finds. While sex role training begins early in life, its effects are not seen until parenting. When the chronic emergency for parents is concluded, "contrasexual potentials" emerge. After the child is on his way, in the post-parental years, older parents "finish paying their species dues." They are no longer burdened by "the psychological tax that is levied on our species in compensation for human freedom from the programmed rigidities of instinct" (1987:202). Then, Gutmann concludes, a significant turn-about occurs: men accommodate to their sensual femininity, women to their aggressive masculinity. In a sense, then, in middle age culture the work of the society in reproducing itself constrains nature; in old age, true nature bursts forth as parents once again are free to utilize the energies that had been tied up in raising children.

Gutmann claims, regardless of the movement of the older man away from a stance of aggressive control, that the ethnographic literature is filled with examples suggesting that in traditional societies, the political and social authority of older men increases.

We should think about this for a moment. Is his attempt to extend what are primarily data from projective tests to the realm of behavior performed only through suggestive assertion? It would appear so. This is the weakest link of his argument.

Two questions comes to mind: if there is a switching of gender potentialities over the life course or in later life, how does one explain both the growth said to occur in the authority of older men and the continuation of male dominance said to occur in later life? What good is this for women? Or for the psychological health of men?

It is important to note, too, that while the increase in male authority may be seen here as problematic, it does not seem to present a problem for Gutmann's argument. It hardly seems possible that such radical changes in gender-linked psychological dispositions should lack real behavioral correlates. Yet dispositional changes appear to be negatively correlated with behavior: it appears that as men age in traditional societies (replacing their aggressive masculine dispositions with feminine ones) they also gain more political and social power and control: "Humans then are the only primate species in which the older male, despite his loss of physical strength, can remain in the human band and find alternate dominance hierarchies wherein status depends on qualities other than brute power, cunning, and ferocity" (1987:215).

He continues, "In the human case, institutions that are as unique to our species as male gerontocracy itself may underwrite the physical, psychological, and social survival of the aging man, at least until he can gain his own secure place as a sacrosanct male elder. The incest taboo, for example, guarantees him sexual access to at least one female, his wife, for who he does not have to fight" (1987:216) He notes, too, "apparently, the androgynous older man is above the battle; he does not stir the sense of threat in young men, and there is no honor and much guilt to gained from killing him" (1987:216).

This is a mixture of generalizations and assertions quite rivaling 19th century thought about 'the mind of the primitive.' The idea that the male elder is above the battle is one of a number of solutions to the paradox of blending of psychic interiority and passive mastery with the increase in political authority that Gutmann discusses. Other solutions that he mentions are: that younger and older men are directed towards different goals and careers; that older men are less apt to control resources of economic production but, in contrast, show increased ritual power; that there is an association of the "sacred" system with the prestige of the elder male so that older men gain spiritual status because of their prominent passivity. In Gutmann's terms, the old man becomes a hero, a "culture tender" and plays as well a secondary but vital parental role.

There is another possible alternative here as well, one that Gutmann fails to consider at all, namely that the description of traditional societies as gerontocracies on the Western model is inaccurate and inadequate. Is the existence of so-called male gerontocracies as common (universal) as he assumes?

There can be two views (at least) here. The first suggests that gerontocracies of a sort may exist, incorporating structures of hierarchy and sexual asymmetry, but these must be understood not as timeless social forms ("a Gerontocracy") but as the product of particular historical forces and power circumstances. A second approach suggests that gerontocracies do not exist and are the object of misrepresentation through male bias, "hegemonic" thinking, and a failure to understand forms of female power. In this vein, commonly recognized forms of sexual asymmetry in society (such as our "public" and "private" and the domestic and politico-jural domains), that have come under attack by feminist scholars, are also to be criticized as they apply to a political "domain" structured on the basis of age (see below). Thus, from this point of view, it is not merely that much data in anthropology have been gathered through the lens of male bias, but also, under the influence of feminist anthropology, we are becoming increasingly aware of the diversity of culturally constituted powers, resources, valuables and domains over which women may have control or influence. Gutmann completely ignores recent scholarship in this area (for example, see Collier and Yanagisako 1987).

Indeed, there would be certain advantages to be gained from Gutmann's "post-parental" perspective by adopting a critical view of the existence of gerontocracies. Perhaps the most salient advantage is that the difficult jump from gender disposition reversal to a glaringly mismatched gender-based political structure would not have to be made.

Gutmann's view may be described as culturalized in its commitment to making use of a wide variety of ethnographic data. But it does so uncritically, replacing native meanings with those supplied by the outsider. Moving from culturalized work to fully cultural work would require the argument to make a further investment in coming to understand the cultural meanings (ours as well as other people's) of notions such as parenthood, which are used rather uncritically by Gutmann.

I have dwelled on Gutmann's argument at some length because it seems to me to be an extraordinarily significant one particularly in raising, as it does, the question of aging in terms of parenting and in species and evolutionary terms. And certainly, insofar as data are provided on cultures with which he has first-hand familiarity (such as the results of projective tests), his argument is partially documented. But as this material begins to be situated in an expanding theoretical argument, greater difficulty is encountered (some problems include lack of depth in ethnographic citation – for example data on the hunger of older men ignores a wide range of data on food cross culturally), a habitual use of ethnocentric categories, and wide ranging projection of ideas onto data that will not support them.

However, these reservations and the question of role reversals aside, there are still several difficulties. Like many who argue about species and evolution on a cross cultural basis, there is a focus here on putative similarities at the expense of differences. While for example he does discuss the effects of modernization and urbanization on elders, he seems to suggest that life for older males is quite good in culturally pristine, traditional settings, falling captive then to the "myth of the Golden Isles" that Nydegger (1983) has described to illustrate the widespread, but incorrect, ethnocentric notion that the life of elders must be better elsewhere. Similarly, the work of Glascock (1982), documenting a relatively high frequency of poor treatment for elders in traditional societies is ignored.

There is too a neglect of the role of men as parents and the culturally defined contributions of men in the making and nurturing of children. There is the spurious notion that men are warriors in all societies. For example,

Bernardi, in his careful review of age class systems notes, "Age class systems have often been described as bearing such a close relationship with military activity as to be entirely identified with the organization of warriors....The attribution of an exclusively military nature to age class systems is a distortion" (1985: 30-1). And most significantly, the cultural meaning of gender is not treated.

Again, I refer to Gutmann's treatment of gender and age as "culturalized" because while there is an admirable interest in treating questions of culture, there is a failure to break with the heavily laden concepts that are part of our own Western operating culture and our scientific culture as well. Ethnographic sources are not read critically. Gender roles are stereotyped and there is little attention paid to diversity. If we begin with the premise that cultural meaning is fully arbitrary, we take a step back from and avoid these sorts of problems, and we begin to look at a wider range of possibilities for "age" and "gender."

While Gutmann's project is of one sort, a related set of problems encumber another significant area of cross-cultural aging research. Consider the role of gender in the study of age class systems which "through the classifications of coevals into formal social groupings....provides for the promotion of all members of society along a scale of social gradation" (Bernardi 1985: xiii). In such systems, "age assumes a pre-eminent role as a principle of social structure" (1985:2). Widely distributed, age class systems may conceptually blend with systems of male initiation, secret societies, social marking of maturation. Bernardi suggests that in stateless states, age class systems are one of five sorts of political systems that may be found. He views the age class systems as fundamentally political. In his discussion of these structures, information on women is intentionally "almost entirely" omitted from discussion of the models that represent age classes. In a chapter devoted to women in age class systems, in general data on women is found to be poor. In 20 East African societies with age classes, three have women's age classes and three have women's age classes that are assimilated into men's classes; in 14, such data is "not mentioned or absent" (cf. Kertzer and Madison 1981). And even in those societies listed as having age grades, the data are ambiguous and equivocal. For the one society for whom there is reliable data – the Latuka – there are a number of bases of similarity for male and female classes including universality of membership, similarity of grades, equality of class members and the function of the classes in organizing collective labor, there are major differences. In women's classes, there is less formalization of passage from class to class; they are more individualistic; adolescent participants are recruited by men, at a younger age than men; there is less similarity in the structure of women's classes from village to village and most significantly from a political perspective, there is political correspondence: male adults are "owners" of the village; female adults do not have control of recruiting or commanding girls and "have no part in the exercise of political and juridical power" (1985: 137). And in a brief review of matrilineally and bilaterally organized African societies with age classes, Bernardi notes, too, that the female age classes lack the formalism and power of the male classes. He concludes (1985:141) that "age class system have a decisively masculine character" and the position of women is defined in relation to the male systems. This conclusion reaffirms Bernardi's sense that the purpose of these systems is "political."

This is a conflation of age and gender that merits comment. Bernardi indicates that ethnographic knowledge of age classes, and their gender aspects, is uneven at best, poor by and large. If it is more important for men to mark "age," can it be said that women age in the same way as men? Suppose gender asymmetry represents some reality. A basic question is, How are we to account for this asymmetry culturally (if indeed gender differences are not hinged to some "innate" difference?)

Even attempts to make radical the culturalized approach, yet failing to make a clean break with our cultural constructs, are problematic. For example, Gailey (1987) argues that gender must be distinguished from biological sex and that it is theoretically possible to acknowledge (at least) four genders, children (for whom biological reproduction is not relevant), adult males and females of reproductive age, and elders who "occupy a status defined by age rather than reproductive capacity." This is an interesting idea but it ignores first, the potential of older men as human reproducers, and second and more significantly, the totality of all cultural things that are or might be reproduced by individuals at certain ages or regardless of age.

*The Cultural Approach*

This brings us to a discussion of the cultural approach. It is important to note that there is no intrinsic reason that the associations of gender and aging fall out in any particular way. This approach suggests that the meaning of gender and age be more purely situated in the realm of cultural meaning and not in any "intrinsic" characteristics we may perceive in them on the basis of our own cultural meanings (i.e., "nature").

In order to begin to view "age" more fully culturally, it is necessary to examine developments in kinship and gender theory that themselves have begun to push the study of kinship beyond the limitations of our own folk biological models (upon which our "scientific" study of kinship, by means of both the genealogical method and our everyday kinship reckoning have been based). I will rely here on an essay by Collier and Yanagisako (1987) that seeks a unified theory of gender and kinship, and I will use some of their remarks for framing questions about age.

Collier and Yanagisako make several important points that speak directly to our attempt to free our view of "old age" as based on "the biological facts of aging." In their essay they challenge the status of "kinship" as ultimately based on "the facts" of sexual reproduction. Recognizing that sexual inequality, once defined as a "natural" fact, is now viewed as a "social fact," they find this transformation as not yet sufficient to liberate gender and kinship studies from their bases in our folk model of biology and naturalness and render them truly scientific and culturally sensitive. Central here is the question of the *difference* between male and female (a difference which itself is unquestioned in anthropology and presumed to be universal.) They ask what are the specific social and cultural processes that cause the genders to *appear* different and whether these differences are sufficient to merit the universal existence of "male" and "female" as cultural categories?

In opposing the notion that cross-cultural variation in gender definitions are merely transformations of a basic "natural" category, they urge that we go beyond our own folk models. Our model bases the existence of *both* the domain of kinship and the domain of gender on "the facts of sexual reproduction." This, they suggest, is not universal or universally relevant.

These ideas are germane to the task of making the study of aging truly cultural. In academic culture, and to a certain extent in Western society at large, no longer are the "declines of age" fully viewed as "natural." Many are now identified as social or socially produced. There is an increasing recognition, for example, that disease is not part of "normal" aging and that physical decrements are not impediments to expressive personhood. Yet while we are aware of the cultural potential for great age to bring honor, not shame, our view of age has much to do with our biological thinking and our chronic sense of temporal linearity. We have trouble freeing ourselves for a truly cultural view of "aging." We have difficulty dealing with perceived differences, the irresolvable gap, between old age and youth.

It is quite possible that either through the perspective of generation or through the ideological dominance of folk biology in our thinking, the "unified analysis" of kinship and gender can be enlarged to include both generation and age.

A more culturally situated approach is taken by Counts and Counts (1985) in their significant overview of aging in Pacific cultures. While it is a gerontological truism that the effects of aging and gender are intertwined, research in Melanesia – the area of the world that includes New Guinea, the Solomon Islands, and Vanuatu – has shown how, culturally speaking, "gender and age are interlocked in a transformational process that unfolds as the individual moves through the life cycle" (1985:8). These findings are significant and bear restatement here: (1) some New Guinea peoples view gender as mutable and expect changes to occur as people age; (2) humans may be culturally viewed as inherently androgynous or naturally female so that being male needs to be worked at; (3) gender is not fixed and may change at any time during the life cycle including old age (Counts and Counts 1985: 8-10). Similarly, they note the widespread distinction made in many societies between those elders who are decrepit and those who can still perform significant cultural activities. They cite examples from several Pacific languages which associate "old people" with such ideas as 'dead', 'corpse', 'ill', 'wretched', 'powerful', or 'ruined' (1985:12).

Illustrative here are accounts by Meigs (1976, 1984) of the Hua of Papua New Guinea. This culture, like many others in Melanesia, features a focus on gender asymmetry and widespread concern with the effects of pollution, through contact with deleterious bodily substances, that may effect one's purity and health. Among the Hua, men imitate menstruation, men believe they can become pregnant, and post-menopausal women are initiated into the men's society and "take on the ritual status of male vulnerability" (1976:394), that is, are susceptible to the inimical effects of pollution. Old men become more like women, lose their vulnerability, are able to eat foods that once might have polluted them, and "avoid nothing." On the basis of the primary dichotomy that structures Hua society, the struggle for purity and the fear of pollution, polluted people – polluted social categories – include for various reasons: women of reproductive age, old men, children, and post-menopausal women with less than three children. Among the Hua, the meaning of age is clearly situated in the context of a world view that focuses on gender, reproduction, and pollution.

For us, kinship and gender are about "the facts of biological reproduction." For others, these may be based on some distinctive principle or principles. For us, kinship is rooted in the biological because in our own definition "it is about relationships based in sexual reproduction" (Collier and Yanagisako 1987:30). For us, "aging" is rooted in biology because it is about "generations" (these too based in sexual reproduction and the genealogical grid) and it is about the representation of chronological time, certainly one of the most significant and pervasive of Western cognitive structures.

Thus, assumptions about the natural entailments of biology and linear time lie at the basis of our own cultural and scientific views of "aging" and "old age." An attempt to study aging and old age cross-culturally without accounting for the effects of our notions of these things and without a serious look at other people's notions, will be troubled.

The dilemma of our own folk model is that it reduces anything without a social or a cultural explanation to a natural or biological explanation. A way out of this conceptual dilemma is offered by Collier and Yanagisako. One must step back from engagement with assumptions, they argue, and one must analyze social wholes, particularly in terms of history and dimensions of cultural meaning. One must *a priori* assume, too, that social systems are systems of inequality, but the nature of this is to be specified locally, not assumed. One must examine the symbol system, freed from assumptions, in terms of the system of social action.

These comments of course apply with equal precision to age and aging. I believe that it is particularly important that ethnographers pay attention to age in the context of the overall system of inequality (Rubinstein

1986b) and the totality of culturally defined values to which individuals have access. The focus on cultural meaning, unencumbered by our own assumptions, is critical as well.

We know what we mean by aging and old age. We have no trouble identifying older people here, and anthropologists do not have this problem elsewhere. It is alleged that the category of old person is universal. But if we really believe in the relativity of cultural meaning, and are willing to let others' meanings stand independent of our own, we should be ready for the cultural reality that somewhere "old age" may not correspond completely, partially, or even at all, to our notions of it.

## REFERENCES

Alexander, J. (1986) Women and Aging: An Anthology by Women. Corevallis, OR: Calyx.

Arens, D. A. (1982) Widowhood and Well-being: An Examination of Sex Differences within a Causal Model. International Journal of Aging and Human Development 15: 27-40.

Arluke, A. and J. Levin. (1985) "Second childhood": Old Age in Popular Culture. In Growing Old in America. B. Hess and E. W. Markson, eds. (Third Edition). New Brunswick: Transaction Books.

Atchley, R. C. (1975) Dimensions of Widowhood in Later Life. Gerontologist 15: 176-178.

Beeson, D. (1975) Women in Studies of Aging: A Critique and Suggestion. Social Problems 23: 52-59.

Bell, M. J. (1986) Women as Elders: Images, Visions and Issues. New York: Haworth.

Berardo, F. M. (1968) Widowhood Status in the United States: Perspectives on a Neglected Aspect of the Family Life Cycle. Family Co-ordinator 19: 11-25.

Berardo, F. M. (1970) Survivorship and Social Isolation: The Case of the Aged Widower. Family Co-ordinator 19: 11-25.

Bernardi, B. (1985) Age Class Systems: Social Institutions and Polities Based on Age. Cambridge: Cambridge University Press.

Bleier, R. (1984) Science and Gender: A Critique of Biology and its Theories on Women. New York: Pergammon.

Block, M. R., J. L. Davidson, and J. D. Grambs. (1981) Women over Forty: Visions and Realities. New York: Springer.

Brown, J. (1985) In Her Prime: A New View of Middle-aged Women. South Hadley, MA: Bergin and Garvey.

Collier, J. F. and S. J. Yanagisako (1987) Gender and Kinship: Essays Toward a Unified Analysis. Stanford: Stanford University Press.

Counts, D. A. and D. R. Counts (1985) Introduction: Linking Concepts Aging and Gender, Aging and Death. In Aging and Its Transformations. D. A. Counts and D. R. Counts, eds. Lanham, MD: University Press of America (Association of Social Anthropology Monograph Number 10).

Farrell, M. P. and S. D. Rosenberg (1981) Men at Midlife. Boston: Auburn House.

Fausto-Sterling, A. (1985) Myths of Gender: Biological Theories about Women and Men. New York: Basic Books.

Finch, J. and D. Groves (1983) A Labour of Love: Women, Work and Caring. London: Routledge and Kegan Paul.

Formanek, R. and A. Gurian. (1987) Women and Depression: A Lifespan Perspective. New York: Springer.

Fortes, M. (1984) Age, Generation, and Social Structure. In Age and Anthropologfical Theory. D. I. Kertzer and J. Keith, eds. Ithaca: Cornell University Press.

Gallagher, D. E., J. N. Breckinridge, L. W. Thompson, and J. A. Peterson, (1983) Effects of Bereavement on Indicators of Mental Health in Elderly Widows and Widowers. Journal of Gerontology 38: 565-571.

Gee, E. M. and M. M. Kimball (1987) Women and Aging. Toronto: Butterworths.

Gibson, M. J. (1985) Older Women around the World. Washington DC: International Federation on Aging.

Glascock, A. (1982) Decrepitude and Death Hastening: The Nature of Old Age in the Third World. In Aging and the Aged in the Third World. J. Sokolovsky, ed. (Studies in Third World Societies, Number 23).

Goodale, J. (1980) Gender, Sexuality and Marriage: A Kaulong Model of Nature and Culture. In Nature, Culture and Gender. C. MacCormack and M. Strathern, eds. Cambridge: Cambridge University Press.

Gutmann, D. (1987) Reclaimed Powers: Towards a New Psychology of Men and Women in Later

Life. New York: Basic Books.
Harris, J. (1975) The Prime of Ms. America: The American Woman at Forty. New York: Putnam.
Haug, M. R., A. B. Ford, and M. Sheafor (1985) The Physical and Mental Health of Aged Women. New York: Springer.
Hawley, D. L. (1985) Women and Aging: A Comprehensive Bibliography. Burnaby BC: Simon Fraser University, Gerontological Research Centre.
Helsing, K. J., M. Szklo, and G. Comstock (1981) Factors Associated with Mortality after Widowhood. American Journal of Public Health 71: 802-809.
Henig, R. M. (1985) How a Woman Ages. New York: Ballantine Books.
Heyman, D. K. and D. Gianturco (1973) Long-term Adaptation by the Elderly to Bereavement. Journal of Gerontology 28: 350-353.
Hyman, H. H. (1983) Of Time and Widowhood: Nationwide Studies of Enduring Effects. Durham NC: Duke University Press Policy Studies.
Kendig, H. L., R. Coles, Y. Pittelkow, S. and Wilson (1987) Confidantes and Family Structure in Old Age. Journal of Gerontology 43: S31-S40.
Kertzer, D. I and O. B. B. Madison (1981) Women's Age Set Systems in Africa: The Latuka of southern Sudan. In Aging, Culture and Health. C. L. Fry, ed. New York: Praeger (Bergin).
Levinson, D. J. (1978) Seasons of a Man's Life. New York: Knopf.
Lewis, R. A. and R. Salt (1986) Men in Families. Beverly Hills, CA: Sage.
Lopata, H. Z. (1979) Women as Widows. New York: Elsevier.
Luborsky, M. (1987) Analysis of Multiple Life History Narratives. Ethos 15: 366-381.
MacCormack, C. (1980). Nature, Culture, and Gender: A Critique. In Nature, Culture and Gender. C. MacCormack and M. Strathern, eds. Cambridge: Cambridge University Press.
McMorrow, F. (1974) Midolescence: The Dangerous Years. New York: Quadrangle.
Meigs, A. (1976) Male Pregnancy and the Reduction of Sexual Opposition in a New Guinea Highlands Society. Ethnology 15: 393-407.
Meigs, A. (1984) Food, Sex and Pollution: A New Guinea Religion. New Brunswick: Rutgers University Press.
Melamed, E. (1983) Mirror, Mirror: The Terror of Not Being Young. New York: Linden Press.
Morgan, L. (1976) A Re-examination of Widowhood and Morale. Journal of Gerontology 31: 687-695.
Myerhoff, B. (1984). Rites and Signs of Ripening: The Intertwining of Ritual, Time and Growing Older. In Age and Anthropological Theory. D. Kertzer and J. Keith, eds. Ithaca: Cornell University Press.
Nett, E. M. (1982) Women as Elders. Toronto: Resources for Feminist Research.
Nudel, A. (1978) For the Woman over Fifty. New York: Taplinger.
Nydegger, C. (1983) Family Ties of the Aged in Cross-cultural Perspective. Gerontologist 23:26-32.
Ortner, S. (1974) Is Female to Male as Nature is to Culture? In Women, Culture and Society. M. Z. Rosaldo and L. Lamphere, eds. Stanford: Stanford University Press.
Ortner, S. B. and H. Whitehead. (1981) Introduction: Accounting for Sexual Meanings. In Sexual Meanings: The Cultural Construction of Gender and Sexuality. S. B. Ortner and H. Whitehead, eds. Cambridge: Cambridge University Press.
Ostor, A. (1984) Chronology, Category, and Ritual. In Age and Anthropological Theory. D. Kertzer and J. Keith, eds. Ithaca: Cornell University Press.
Pesman, C. (1984) How a Man Ages. New York: Ballantine Books.
Rosaldo, M Z. (1987) Moral/analytic Dilemmas Posed by the Intersection of Feminism and Social Science. In Interpretive Social Science: A Second Look. P. Rabinow and W. M. Sullivan, eds. Berkeley: University of California Press.
Rubinstein, R. L. (1986a) Singular Paths: Old Men Living Alone. New York: Columbia University Press.
Rubinstein, R. L. (1986b) What is Social Integration in Small Scale Society? The Journal of Cross-Cultural Gerontology 1:391-409.

Sahlins, M. D. (1976) The Use and Abuse of Biology: An Anthropological Critique of Sociobiology. Ann Arbor: University of Michigan Press.

Sankar, A. (1984) "It's Just Old Age": Old Age as a Diagnosis in American and Chinese Medicine. In Age and Anthropological Theory. D. Kertzer and J. Keith, eds. Ithaca: Cornell University Press.

Schneider, D. (1972) What is Kinship All About? In Kinship Studies in the Morgan Centennial Year. P. Reining, ed. Washington DC: The Anthropological Society of Washington.

Szinovacz, M. (1982) Women's Retirement: Policy Implications of Recent Research. Beverly Hills: Sage.

Shapiro, J. (1981) Anthropology and the Study of Gender. Soundings: An Interdisciplinary Journal 64; 446-65.

Silverman, P. R. (1981) Helping Women Cope with Grief. Beverly Hills: Sage.

Simon, B. L. (1987) Never Married Women. Philadelphia: Temple University Press.

Siskind, J. (1978) Kinship and Mode of Production. American Anthropologist 80: 860-872.

Strathern, M. (1980) No Nature, No Culture: The Hagen Case. In Nature, Culture and Gender. C. MacCormack and M. Strathern, eds. Cambridge: Cambridge University Press.

Walker, B. G. (1985) The Crone: Women of Age, Wisdom and Power. San Francisco: Harper and Row.

Young, M., B. Benjamin, and C. Wallis (1964) The Mortality of Widowers. Lancet 2: 454-456.

CHRISTINE L. FRY

# 5. THE LIFE COURSE IN CONTEXT: IMPLICATIONS OF COMPARATIVE RESEARCH

INTRODUCTION

As age and gerontology became prominent research issues, lives have become central researchable units. The life span or life course perspective has emerged along with gerontology and has invaded most of the social and behavioral sciences. The genesis of this perspective stems from the recognition of two theoretically salient issues. First, for those studying the last stages of life, we cannot divorce them from the age grades which precede them. Secondly, for those researching aging, the conclusion is, it isn't age, per se, that is the object of study. We need to examine lives in their entirety and we need a unit through which we can generalize about lives and social organization. This unit is the life course.

Early examination of human development centered on childhood and concluded with the finished product of adults. When the crisis of the Great Depression combined with the increasing presence of older adults, gerontology emerged to focus on the processes and problems of the last stages of life. With either side of social maturity charted, adulthood appeared as a monotonous plateau extending from puberty to senescence. On the other hand, it is rather obvious that the entrance into old age did not pivot on one event: retirement, widowhood or even the 65th birthday. Nor did becoming old transform an individual into someone remarkably different. Thus, continuity with the earlier stages of life became a productive strategy in understanding the processes and diversity of pathways in growing old.

Age, conceptualized chronologically, also proved to be a problematic variable in research on aging. "Age obfuscates more than it elucidates!" is a frequent lament. The reason is that age is a proxy for other things: social maturity, eligibility, or even need (Neugarten 1982). In large scale, highly differentiated, bureaucratic societies people need shortcuts to simplify the social diversity of a nearly infinite social field. Others are placed in categories – "the old," "the over 60's," or "the late 70's" which become reified into uniform groups. As scientists we know better than that. Consequently there has been a call for more refinement in our conceptualization. A part of this refinement was the recognition that aging occurs on several levels simultaneously. At minimum these are biological and social. Chronology (time elapsed since birth) does not predict biological aging because of diversity in genetic endowments and environmental influences. It is even less predictive of social aging where such factors as gender, class, race, and ethnicity produce even more diversity along with choices individuals make as they move into and through adulthood. Conceptually, the life course is a more attractive variable than chronological age simply because it increases the validity of the phenomena we study.

The purpose of this chapter is to examine the life course as a researchable unit, using the comparative and cultural perspective of anthropology. First we will consider the origins of the life course perspective, the assumptions made by this approach, and research agendas stimulated by this paradigm. Secondly we will examine the implication of the comparative perspective for the life course as a unit and as a universal pattern for comparison across cultures. Finally we will turn our attention to models of social organization and the theoretical implications of the life course.

## ORIGINS OF LIFE COURSE PERSPECTIVE

In searching for the roots of the life course perspective, we find they are the product of many lines of inquiry both social and behavioral. Social sciences don't usually focus on the individual, but on social facts. However, in the early part of this century, life histories proved to be a fruitful and exciting source of data. America was experiencing change with incoming waves of immigrants. Lives beginning in Europe and continuing in America could provide insight into that change. Thomas and Znaniecki's *Polish Peasant in Europe and America* (1918-20) revealed the discontinuities in immigrants lives. Similarly anthropologists were experimenting with a genre of "ethnographic biography" (Dyk 1938; Lafarge 1929; Radin 1920, 1926; Simmons 1942). Culture's impact upon individuals could be unraveled as well as their experiences of a changing world.

Although attempts were made to codify and systematize the use of personal documents in social science research (Allport 1942; Dollard 1935; Gottschalk, Kluckhohn and Angell 1945), the post World War II era saw an abandonment of life histories as important data sources. The priorities and paradigms of sociology and anthropology had changed. It wasn't until the 1960's that the interest in life histories and individual lives was rekindled (Bertaux 1981; Langness 1965; Langness and Frank 1981).

Behavioral sciences also focus on change, although typically the emphasis is on change within individuals over time. Indeed, the earliest and major proponents for a life-span perspective have been within psychology. Just as life histories flourished in sociology and anthropology, psychoanalytic psychologists used biography to chart adulthood (Buhler 1935, Kardiner 1945) and to extend the stages of childhood into adulthood (Erickson 1959). The history of developmental psychology is rich (see Reinert 1979 for a review) extending back to the classical philosophers. In spite of such an early beginning, it wasn't until after World War II that a life span perspective began to become differentiated from child dominated developmental psychology. Beginning in the 1970's a series of seminars at the University of West Virginia promoted conceptual and methodological breakthroughs in this paradigm shift (e.g., Goulet and Baltes 1970; Baltes and Schaie 1973; Datan and Ginsberg 1975; Callahan and McClusky 1983).

Although the life span perspective revolutionized psychology, gerontology was its birthplace. Questions about aging require antecedents from senescence back to adolescence and even earlier. After the war, the prevalent theories about aging were dominated by biological models. Child-based development paradigms saw an end point in development, maturation. The attempt was to discover a universal process of development and aging. However, when applied to adulthood, these models were soon revised. Instead of a stable end point of senescence, adults become increasingly different as they get older. To explain this increasing variability, social and contextual factors had to be considered as well. Thus a true interdisciplinary merging of interest occurred as social scientists joined developmental psychologists in forging a life-span view of aging.

At the same time sociological and anthropological interest in life histories declined and never congealed into a distinctive life-span paradigm. Instead, these disciplines drew from their strengths in studies of socialization, stratification, norms, roles and statuses, age differentiation, demography and the life cycle. From these strengths, two lines of paradigm building have contributed to the formation of a sociological and cultural view of the life course. The first of these is the work of Bernice Neugarten and her students, beginning in the 1960's, on age norms and the sociocultural structuring of age. The second is the age stratification model developed by Matilda White Riley and her associates in the 1970's.

In 1968 the publication of *Middle-Age and Aging* (Neugarten 1968) marked the emergence of a social-psychological and cultural perspective on aging. Partially inspired by anthropological work on formal age set systems, the important dimensions for this perspective are concepts of age norms and age statuses (Neugarten, Moore and Lowie 1965; Neugarten and Moore 1968). Each society has age norms defining a "social clock" which regulates acceptable age-linked behavior. Either through sanctions or socialization or both, this cultural timetable guides the approximate progression of individuals through role sequences defining age status systems of that society.

Foundation work for the age stratification model began in the 1960's and was synthesized with the publication of *Age and Society: A Sociology of Age Stratification* in 1972 (Riley, Johnson and Foner 1972). Cornerstones for this perspective are derived from the demographic concept of cohorts and from stratification theory. Infants are born providing each society with its main source of membership. Individuals born around the same time form a cohort. In addition to age, each cohort has distinctive characteristics including such things as size and opportunities and conditions shaped by historical circumstance. Cohorts flowing through a society constitute the age strata with the individuals within the respective strata engaging in a complex set of roles. Change in any of the components of this model (cohorts, historical circumstance, age strata) affects the experience of aging within a society.

As the life-span perspective emerged, time became a central concern. At issue is not how time is conceptualized between birth and death. Rather, in documenting change within that span, there are a range of variables which are linked to how we measure time. Because we have only one span of time, when we have time-linked variables, we have confounded variables. "It is as if we used a triangular ruler, each side in a different metric: the faces show different intervals but, since the time is the same, the correlation is perfect. Any two sides yield the third" (Nydegger 1981). Age/period/cohort is one such confound. If one knows age (years an individual has lived), and period (time of measurement), then one can predict cohort (date of birth). Analytically, such logical confounds are difficult to untangle (Schaie 1965; Glenn 1976; Rodgers 1982).

Likewise, when we bring social and cultural factors to the forefront, temporal issues intensify. Life time, historical time and social time (Neugarten and Datan 1973) form the three sided ruler. Life time (age & biological maturation), historical time (major social system events), and cohorts entering society are processes occurring in the same temporal interval. Social time (age norms and age statuses) structures and gives meaning to lives through normative expectations about life stages and allocation of roles by age/cohort. Precisely because social clocks are social and cultural, they respond and are re-negotiated as societal events become history and demographic patterns shift.

Life courses are but yet another "window" through which to view human development and social organization. Temporal processes are a basic part of this window. Lives and societies are forged in time. Individuals mature and change from birth to death. In the same span of time the societies in which they live experience change in some dimensions and in others continuity. Although confounds may be logically unresolvable, it is the awareness of them which is a strength of the life-span perspective. As a result the life course perspectives makes important contributions by enabling us to link societies and individuals in time; to address questions involving different levels of phenomena (biological, psychological, and sociocultural); and to investigate the antecedents to a variety of outcomes. With further development of the life course as a framework within gerontology, its key premises have been formalized.

## PREMISES OF THE LIFE-SPAN PERSPECTIVE

Among the researchers adopting the life-span perspective, there is remarkable agreement on basic premises (Riley 1979; Featherman 1981; Baltes and Nesselroade 1979; Campbell, Abolafia and Maddox 1985). Briefly these are as follows:

1. *Entirety of the life course.* When considering aging as a biological or social process, the entire life course must be taken into account. Aging is not limited to any particular time of life or life stage. It occurs from birth to death.
2. *Multidimensionality.* Aging is multidimensional encompassing interdependent biological, psychological, and social processes. No one of these dimensions is isolated from the other two or independent and causative of the other two.
3. *Context-linked.* The life course for any particular individual or cohort is linked to a specific social, environmental and historical context. This context produces multiple determinants for change or stability for the life course. Its effects are also interactive and cumulative.
4. *Reciprocal effects.* The relationship between individuals and society is reciprocal. Societal norms and role structures set constraints on individuals, but the collective experience of aging of cohorts can change these norms and roles.
5. *Increasing variation with age.* Developmental changes are more normative in the early phases of the life course. However, when applied to adulthood and old age, variation increases. With increasing age, the variation increases within cohorts.
6. *Plasticity.* Behavior and personality are malleable throughout the life course. Intervention efforts are effective in older as well as in younger age groups.

The life course perspective is a model of aging. As a paradigm, it is not a set of propositions about aging which are empirically testable. Paradigms dictate what are important research questions. They tell us what are the salient variables. They also set standards for the adequacy of our data and analytic strategies to evaluate hypotheses or models about the life course. In the following sections of this chapter we examine the research agenda stimulated by this paradigm and the implications of comparative research for the life course perspective.

## THE LIFE COURSE AND RESEARCH AGENDAS

The life course perspective offers us a grand scheme through which to view the elements of the aging process. We see biological individuals flowing through a social system together in cohorts in a historical and cultural context. As they mature as organisms and change psychologically, they are enculturated and socialized into an age status system which is subject to change by the cohorts and by environment and history. It offers any researcher a very large agenda to explore. It is so grand that in looking at the forest, we can easily lose site of the variables. With the exception of the experience of aging, a researcher is like the proverbial child in the candy shop. We have an embarrassing overflowing cornucopia to choose from: different kinds of phenomena; the whole of life;

different contexts and aspects; and change through time.

In spite of lack of theoretical priorities, the life course perspective has flourished. Research and publication has expanded. Within psychology alone, a review of over 4,000 articles on the life course, reveals a lack of an integrated perspective (Birren, Cunningham and Yamamoto 1983). Instead clusters of special interests defy theoretical integration. The same is true in the social sciences. Although this state of affairs is lamentable for paradigm synthesizers, it is reflective of the recency of the framework. Like an evolutionary "adaptive radiation," diverse issues, new and old, are examined in a different light. As hypotheses are evaluated and knowledge accumulates, the scientific cannon of parsimony will call for theoretical integration.

One area where strategies are divided is in methodology and research design. As mentioned above, a major problem has been the confounds of time in unraveling the effects of age/period/cohort on the experiences of aging; a variety of tactics have emerged. At one extreme are attempts to try to strip context and cohort from individuals to examine the effects of age processes on lives. At the other extreme are strategies which maximize the effect of temporal and contextual variables on the patterns of lives.

If a research design is formulated to investigate a law-like process or regularity that is attributable to all people, then the confounds of time and context must be controlled for, or the effects known. One approach has been to use longitudinal research designs. Although longitudinal studies differ in the specifics of their design, by following individuals over time, intra-individual change is measured with period and cohort effects being controlled. Cohort effects are being maximized unless more than one cohort is in the study. A number of longitudinal data sets have constituted an empirical basis for generalizing about change over a life time. These include the Berkeley Longitudinal Studies, the Duke Longitudinal Studies, and the Baltimore Longitudinal Studies as well as others (Eichorn *et al.* 1981; Busse and Maddox 1985; Shock, Greulich and Andres 1984).

Longitudinal designs cannot resolve all the problems (see Maddox and Campbell 1985 for a review and discussion of statistical issues). Although increasingly popular, longitudinal designs are expensive, subject to biases because of attrition, and face the same problems that confront cross-sectional designs in examining cross-individual differences. Additional problems include generalizability and validity to other populations, contexts and cohorts.

At the other end of the continuum of strategies, are designs which maximize the effects of temporal or contextual variables. Historical change within one society is a source of contextual variation. One example of a historical approach to the life course is the work of Tamara Hareven (1982) investigating the effects of industrialization on the family cycle in a New Hampshire mill town. Another example, using archival data from the Oakland Growth Study, is Glenn Elder (1974) who examined the effects of the Great Depression on the adult life course of children who experienced economic deprivation as adolescents. Other designs for examining the life course are comparative, selecting communities, ethnic groups, or societies and cultures precisely because of their sociocultural differences.

Comparisons are at the heart of all research designs ranging from across individuals, across time or across situations or contexts. Cross-cultural designs maximize the variation across contexts included in the study populations. Aging is experienced in specific socio-cultural environments (the context-linked premise discussed above) which include specific norms, values and perceptions about aging. Lives and age are further structured by social institutions (e.g., family, religion, economy, politics) which are variable across contexts.

Two features make cross-cultural designs very attractive. First, by investigating socio-cultural diversity we may ultimately be able to distinguish what is universal about the experience of aging from the contexts in which it occurs. If we minimize and homogenize the diversity, then we may never know what the range of cultural effects are in the real life settings where people are born, mature and die. Secondly, in examining the life course in other cultural contexts, corrections are made to ethnocentric biases that inevitably result from research in one society or societal type. Gerontology is the child of industrialized nations, especially the United States and to a lesser extent other developed nations. Consequently, most of our empirical evidence and theoretical conceptualizations of aging are based on a narrow range of world-wide societal variation. With 75 – 80% of the world's population in underdeveloped countries (Hauser 1976), the rationale for expanding our data base is obvious.

Units for cross-cultural comparisons must be identifiable and operationalizable in different contexts, the parameters of the unit should be demarcated, as well as the possible ways it can vary. It is now to the question of the life course as an unit that we turn.

## THE LIFE COURSE AS A COMPARATIVE UNIT

Recently, there have been calls to consider the life course as a cultural unit (Fry and Keith 1982; Fry 1986). However, comparative researchers, primarily anthropologists, have not embraced the life course perspective and elevated it to a central focus in their research agendas. With very few exceptions (eg. Levine and Levine 1985), the life course has not been a window to cultural diversity. Some progress has been made, however, as anthropologists introduce some of the lessons learned in the multidisciplinary field of gerontology into to the core of anthropology (Keith 1982; Kertzer and Keith 1984). One example is Project AGE which has used the life course as a major part of its study of community effects on the well-being of older people in seven communities around the world (Ikels, Keith and Fry 1988; Fry 1988; Keith, Fry and Ikels 1990).

As a comparative unit, the life course is not, in itself an institution. Institutions usually resolve essential problems; have sets of beliefs and expectations; and structurally arrange personnel in a web of social relations. Becoming, living and dying, present rather basic problems in any sociocultural context, but those problems are resolved in various institutions. For instance reproduction, enculturation, and domestic life are at the core of kinship. Making a living and exchanging and distributing goods and services are the raw material of economics. Likewise power, authority, conflict resolution and decision-making constitute the grist for political institutions. Religion answers the ultimate questions and deals with the supernatural and ritual life. The point is, many basic concerns are worked out within the institutional structure of a society. Some of these concern the life course, others do not. Thus the life course also becomes intertwined other institutional segments such family cycle and in industrialized societies career lines unfold in the occupational structures of the economy.

Besides being interconnected with social institutions, the life course does not become a basis for age-based groups because cohorts rarely form groups with collective identities. The major exception to this are age-set societies from Africa, Australia, Melanesia, South America and the Great Plains of North America. Here males are recruited into a social group on the basis of their age. These are called "age sets", "generation sets", or "age societies." Once formed, the men will pass through their life courses as a group with intra-set relations with peers and inter-set relations with their junior and senior sets. Because age is formalized into an institution, at least for males, these societies are fascinating laboratories in which to examine life course issues (Foner and Kertzer 1978). (For a review of these societies see Baxter and Almagor 1978; Bernardi 1985; Stewart 1977.)

Despite non-institutional status, the life course is institutionalized in all societies. The degree of its explicitness in definition and standardization is one point of variation. Certainly, the age-set societies where public life is given structure by age and corporate groups recruited by age are one extreme. Although not the basis for life-long collectivities, the life course in modern industrialized societies is also highly institutionalized, but in a different way (Meyer 1986). Here national level institutions use age to recruit and to rationalize large populations. In many respects, it is the institutionalization of the life course in both these kinds societies which has shaped our theoretical model of life courses. If we are to use the life course as a comparative unit, we first have to examine what gives it meaning in every culture and what elements lend it some degree of standardization among individuals. Secondly, because the life course is an etic unit, that is one largely defined by outside observers (sociologists, psychologists, and anthropologists), we should be aware of cultural variation in definition. Variation in emic (native points of view) can raise issues which result in refinement of our units into more valid concepts.

As an etic unit, the life course has a number of elements that define it. First, temporality is central. Humans enter time and society at or before birth, mature, and then withdraw. Secondly, the time interval between entry and exit, has cultural definition. People learn what their futures will be like and what they should do. Their perceptions of the life course chart the path or pathways through time and society. These pathways have their parameters giving them structure and meaning. Thirdly, the passage through time involves transitions as milestones are passed; life stages are entered and exited; and one becomes senior, and then old. Finally, time does not just past in a continuum, but is punctuated into life stages or age grades. Each of these elements are examined using what is presently known from the natural laboratory of other cultures.

*Time*

In spite of the importance of time, with very few exceptions (Hendricks and Hendricks 1976; Hendricks and Seltzer 1986), conceptions of time have not been explicitly considered in our formulations of the life course. Yet time is culturally defined and embedded in the logic of language. Time is a human construction. Through the recognition of repetition, humans define it with beginnings and ends. In between, they add a notion of change and rate of change to arrive at a sense of duration between end points of an interval (Leach 1961). Day/night, winter/summer, birth/death are familiar repetitions. Language further shapes how these are expressed. A past, a present and a future are also verb tenses which are by no means universal (Whorf 1964). The hidden impact of language is that these tenses predispose time to be considered in stages as it passes from past to future. An alternative is to see it as a becoming.

Time is relative, but the Euro-American perspective on time has prevailed in gerontology. Euro-American metaphors are linear. In the beginning time starts its flow like a stream to the present and then on into the future. Similarly, the life course is a linear, irreversible progression between birth and death. Displayed between the intersection of the axis of time and age, cohorts are seen to flow on a diagonal to the right (as in Riley 1986: 370, Figure 1). From birth to death appropriately describes the situation for any one cohort or individual.

Comparative data tell us that lines are not the only geometric metaphors for time (Bohannan 1953; Evans-Prichard 1940; Maxwell 1972; Smith 1961). Time can be circular with some kind of reincarnation recycling humans through time from birth to death and back again. This view of time is not uncommon occurring in societies as disparate as the Eskimo (Guemple 1980), some East African age-set societies where the age sets are recycled (eg. Masai), or the more familiar Hindu of India. From birth to re-birth might be a more accurate description of life. Helixes are also metaphors. Lives are seen as a worm slithering in wide arcs spacing the Boran generation sets (Baxter 1978: 157). Even where time is lineal, death may not necessarily end one's involvement in human affairs (Fortes 1959). In many societies in Africa, Asia and in the Pacific ancestors are yet another stage in human development; from birth to ancestor, may be the end points.

Since three kinds of time (life, history & social) are involved in our conceptualization of the life course, time warrants our attention. Of the three kinds of time, historical time and social time are culturally conceived and are relative to cultural context. However, like many other cultural domains (e.g., color), time is not arbitrary and independent of the phenomenal world. Natural rhythms such as seasonality, day to night, new to full moon, birth, maturation, menstruation, death etc. are the raw materials with which cultural conceptions of time must reckon. Gerontology advanced when the life course replaced age as the primary unit of analysis. Likewise, the life course will improve our understanding of aging as folk models of time add to the core element of the etic concept of the life course.

*Perceptions Of The Life Course*

Children generally learn from adults where they have come from and where they are going. What they learn is an anticipation of social time. Also they learn about the pattern of the life course: the path that is possible and expected; or the alternative paths which may be taken. In considering perceptions of the life course, we first must look at the way life paths are conceived in their entirety. Secondly, we must consider the parameters of social time.

Cultural variation in the way the life course is conceived, planned for and evaluated is tremendous for both men and women. Themes and metaphors used to describe life reflect both local circumstance and the degree of integration within the life course. In other words, life can be conceived as having a definite trajectory or it can be thought to just happen in an amalgamation of domains, such as reproduction or spirituality, and seasonal rounds of work. For Tiwi males of Melville Island off of Australia, life had a definite trajectory in marriage negotiations (Hart, Pilling and Goodale 1988). To become a "big man" a Tiwi male had to accumulate many wives. In South America the Quechua Indians of the Peruvian Andes see their lives as an unfolding of vigor and then a decline and loss (Smith 1961). For the long living people of the Caucasus, the Abkhasians, life is filled with moderation and continuities and claims to longevity (Benet 1973). On the other hand, in the Arctic, a harsh environment reinforces a theme of productivity in the lives of Eskimos. Living through the seasons, there is no peak to life, only a gradual curve of increasing respect for age (Smith 1961). "Renewal" strategies allying both older men and women with those who are younger, prolong productivity for as long as possible (Guemple 1969).

Life courses may also vary by the amount of choice and alternatives in the pathways. In urban societies that are highly differentiated in role structure (especially in occupations and voluntary associations), life ways are nearly infinite. Choice and opportunity abound in the many branching paths. Lives are plural, differentiated by class, race and other factors in addition to gender (Hagestad and Neugarten 1985). In less differentiated, smaller scale societies, the life course is more monolithic with gender dividing the alternatives. For instance in Guatemala, the life course in Atchalan is clearly defined for men and women with little or no room for variation and experimentation. On the other hand for Mestizos, living in the same village, life careers are diverse. Contacts, skills and education have linked them with the larger social arena of Guatemala (Moore 1973).

Life courses, regardless of how integrated or fragmented they are, how diverse or monolithic they are, are all structured by social time. How is social time defined? The answer to this question takes us to roles, the building blocks of social analysis. Roles and the attributes they entail (e.g. such things as responsibilities) are markers punctuating the life course and rendering a social clock. Only roles which are age-salient form social clocks. Roles are enacted in a variety of social institutions, most of which are age-irrelevant. For instance being a member of a church tells one very little about life stage because people of all ages belong to churches. On the other hand, being a parent of small children tells a lot more about life course position. The roles which form the workings of social clocks are again culture and context-specific. Before we can begin to compare social clocks, it is necessary to identify the markers of the life course within each setting. Fry (1986) discusses research strategies utilized in identifying emic markers.

Pioneering empirical work on social time was begun by Bernice Neugarten and her colleagues using U. S. respondents. Five dimensions were identified: career lines, health and vigor, family cycle, psychological attributes, and social responsibilities (Neugarten and Peterson 1957). My own work using a different methodology, but again with a U.S. population, revealed that married and family (family cycle), education and work (career lines), and living arrangements were important yardsticks (Fry 1980).

More diversity in attributes punctuating the life course is seen in the communities investigated by Project AGE (Fry 1988; Keith and Fry 1988). This project is a cross-cultural project to investigate the life course and sources of well-being in old age in seven communities around the world (the !Kung and Herero of Botswana, urban Hong Kong, Clifden and Blessington, Ireland and Swarthmore, PA and Momence, IL in the United States). Family (being marriage and having children) is a common dimension across research sites. Similarly, work is another major life course attribute which includes both labor force participation and subsistence. Living arrangements are also societal specific indicators of where people are in the life course. Education is a theme of only developed societies. In the U.S. sites, we find differences. In the more affluent suburb of Swarthmore status of parents is an important marker (still living, working, retired). In the blue collar small town of Momence, housing tenure (mortgage paid or not) marked a major watershed, as is participation in community organizations. In the Irish communities immigration of family is an indicator. In both Irish communities and the U.S. sites, as well as the !Kung, functionality of self and spouse are relevant signs of life stage.

Although some of these dimensions such as work and family are probably universal, the content of the yardstick is culture specific. For instance, even in the U.S. communities, work is conceived differently. In white collar, affluent communities, occupations have career ladders. The indicators of age here are promotions and major job responsibilities. On the other hand, in working class communities, jobs with career ladders are relatively rare. Such markers as working out of town or buying a business reflect age-linked strategies for financial security and autonomy. Where there are no paid/wage earning jobs to be had, such as in Botswana, subsistence activities are work. In these cases, age-salient features may be expressed through ownership of cattle or in specific kinds of subsistence activity (e.g., hunting, craft work, getting water, or collecting firewood). Likewise, the way the family cycle is graded, reflects community and culture. In societies where education is universal, children's maturation is expressed and ranked using the divisions in the local schools (pre-school, elementary, high school, college). On the other hand, where education is not universal, the development of children is expressed in terms of size (e.g., small) or relative age (e.g., older). Where people live long enough to have great-grandchildren, they become markers of age.

For social clocks to work, individuals need not only be in agreement on the calibration, but they also must feel some pressure to conform to expectations. Legal norms are most obvious in their push/pull pressures on ages of marriage, driving, drinking, voting, or eligibility for pensions. Not all age norms, however, are legal and not all age norms involve role occupancy. Awareness of timetables and being "on time" or "off time" are issues people talk about and are frequently choice items of negative gossip as in the case of an "80 year old father." Acting one's age involves conforming to norms of behavior, style, and actions. The initial work on age norms was carried out by Neugarten and her students in Chicago and later replicated both in the U.S. and Japan (Neugarten, Moore and Lowie 1965; Hagestad and Neugarten 1985; Plath and Ikeda 1975). These studies show that perceptions of other people's view about age norms (as compared to one's own view) had more constraint (as measured in age constraint scores) in both Japan and the U.S. and at both times in the U.S. In all three studies as a person aged, their view of age norms showed increased constraint and converges with the perceptions of other people's opinions in the oldest age groups. These studies indicate parallelism in results across time and in two cultures.

Comparative data on norms of role occupation and the impact of age norms on behavior are only beginning to be collected across ethnic groups and cultures. Without these data, it is impossible to compare social clocks and the time they tell.

## Transitions

Roles calibrate social clocks with movement across time signified by role change. Since roles shape the timetables, the life course can be seen as a "role course" (Nydegger 1986a). For a single role we see an entrance, a temporal duration in which that role is occupied, and an exit. The timing of these three components has been a major investigative theme, especially in the study of inter-cohort variation (Demos and Boocock 1978; Hareven 1978; Glick 1977; Hogan 1981; Uhlenberg 1974). Here the role course has been used as a set of variables. Entrance, duration and exit constitute the temporal referents in a rather complex picture. Lives are filled with multiple roles enacted simultaneously in different institutional domains. Some of these roles are decidedly a-temporal or stable having little to do with expectations of age (e.g. church member). Others are clearly temporal involving an ordering with respect to other roles and even sequential progressions of role occupation (e.g., career ladders). Still others reflect the interdependence of roles, as the transitions of linked individuals transform the age status of others (e.g., the domestic cycle where the development of children and their eventual maturation transforms their parents; Nydegger 1986a; Fry 1983).

Although role courses are conceptually attractive, their operationalization yields a complexity producing analytic difficulties. Conceptually, variation in timing of events, synchrony, compressing, normativeness of events, choice in and control over entry, duration and exit are significant variables. First, they can guide descriptions of variability in life courses across time, cohorts, or culture. Secondly, they lead to hypotheses concerning effects on outcomes such as adaptation, stress or later life events.

However, using role courses and social time as tools of analysis present problems. The first is that social time and life time are confounded. Consequently it is difficult to disentangle the age-linked transitions from transitions that culturally define the life course. Secondly, because social time is linked to life time, we may be frustrated when time is not measured chronologically. For instance, in reading ethnographic biography, one is often struck by the lack of age (life time) anchoring of major events (e.g., Shostack 1983). Regardless of its attempt to chart the social events which define progression through the life course, social time is operationally anchored in chronology. Thirdly, social time is calibrated in so far as the salient roles are synchronized, sequenced and ordered. The roles which have received attention combine career and family roles. What happens to social time when there is no career ladder, only work and subsistence? What happens to social time when reproduction takes place in a 30 – 35 year framework and children are not participants in a graded educational system? With chronology and calibration imprecise, social time may vanish as an indicator of age. Fourth, the focus on roles and transitions, may well lead us to miss the combination of factors that people use in making judgments about appropriateness of a transition. These are their theories about the life course which involves more than roles (Nydegger 1986b).

*Age Grading and Age Strata*

Do social clocks produce a continuum of minutes, or organize these smaller temporal units into larger ones? Life is not ordered by minutia, but is cut into larger divisions we call life stages or age grades. Radcliffe-Brown who so long ago (1929) coined the term "age grade" as the recognized divisions of life from infancy to old age, was clarifying a terminological problem in the study of age organizations. For analytic proposes he was differentiating between highly formalized age organizations such as age sets and generation sets and informal age systems. The former have explicit rules in recruiting and establishing social boundaries between groups. The latter, on the other hand, are less precise and are found in all societies approximately grading their members by age or life stage.

Formal age systems, although vastly different than informal age categories, have kindled conceptualizations of age grading. Gerontology, in its infancy, was faced with the question of who are the old? Defining the boundaries for this age grade was both a conceptual and operational problem. Eventually these issues extended to the other life stages of adulthood: What are they? What defines them? Two major strategies have been used. The first theoretically justifies the division of adulthood into etic age stages. These are investigator defined and usually chronologically operationalized. The second, uses emic definitions. Instead of imposing a fixed number of grades with labels such as "young," "middle," or "old," members of study populations are asked to divide adulthood into stages and to give some indication of defining features.

In investigating informal age grading, it should be no surprise that variability is the rule. Lack of precision is a characteristic of informality. Also, the contexts in which this second strategy has been employed also are highly dependent on chronology. Whenever ambiguity occurs in the informal division of life, chronology is invoked as a common and precise denominator. The results we have from three communities in the U.S. (two from Project age and one from the author's previous work) and four neighborhoods in Hong Kong (Fry 1976; Keith, Fry and Ikels 1990) reflect variability. The number of divisions seen range from none to 15, with an average of 5 stages within all research sites. The majority of this variance falls within 3 – 6 life stages. With refinement of divisions varying, the defining features of stages also diverge as the age grades become more differentiated.

If a society is stratified by age, then how do we distinguish the boundaries between layers? Do we use the cultural units of age grades, defined by emics or etics? Emic analysis have the advantage of introducing diversity and local knowledge which produce interesting variables. Etics, on the other hand, are useful, in controlling some of this variability by imposing theoretical direction to the selection of those variables. Another issue is whether cohorts are our unit or are we dealing with something else? Cohorts are temporally defined by historic time. Hence, by using historic time, definition of chronological boundaries is facilitated along with the ability to examine intra and inter-cohort variability. However, the emic and etic polarities raise rather serious problems with cohorts. First their definition reifies chronological age by linking them to historic time and an etic judgment call. As Rosow points out, the culturally meaningful definition of cohort is a problem (Rosow 1978). Since cohorts are emic units defined by historic events and watersheds, if we miss the emic definition by a few years, we are likely to introduce noise into the data which masks the patterns we are after.

## MODELS OF SOCIAL ORGANIZATION AND THE LIFE COURSE

As an organizing framework, the life course perspective has proved to be useful. As a theoretical framework, it has its limitations (Passuth and Bengtson 1988). Part of the limitations stem from the model of social organization that underlies the life course paradigm. Explicitly, it is derived from the Parsonian functional paradigm. Functional paradigms are not especially good for generating hypotheses. What they are good for, however, is telling us what there is to study and the functional linkages between parts. Thus for the life course we see cohorts of humans (biological, psychological, social and cultural beings) entering and exiting time (life time, historical and social time) and society. We see individuals of different ages participating in various institutions perpetuating and changing a society as they mature and leave that society.

Why aren't functional paradigms useful in generating hypotheses? Difficulties with functional models are well-known, but the one which is of interest here is that intra-cultural variation is overlooked. In the search for pattern, variation within units is reduced. Variation across cohorts, historical period, ethnic groups, is sought in order to document the reasons for difference and the reasons for change. Intra-cultural heterogeneity is masked in the search for the normative, the average, and the typical in order to ask the larger questions of cross-group differences.

Along this line an important lesson was learned in the anthropological study of kinship through the debate over descent and alliance (Schneider 1965). Although many issues are involved, the conclusions salient here concern the descent models of lineage based African societies. Anthropologists working in Africa using a Radcliffe-Brown paradigm of functionalism, developed models which examined the world through the window of descent. Descent is everywhere and through ever widening extension of descent (family, lineage, clan, moiety) egos were connected to society and to other descent groups through marriage. Following debates over cross-cousin marriage which involved a number of interpretations and empirical studies (summarized in detail by Buchler and Selby 1968) two conclusions followed. First, descent models did not adequately represent the ethnographic situation even in African societies. In other words, empirically there is much more variation in such issues as post-marital residence, choice of marriage partners and even descent group membership, than the model represents. Secondly, descent models do not work well in many societies – namely, non-African.

Adequacy of representation magnifies some of the difficulties with age norms. A number of investigators have raised issues regarding norms (Hagestad and Neugarten 1985; Ables, Steel and Wise 1980; Dannefer 1988). Much of our understanding of norms and socialization are derived from the work of Parsons. Norms are central in linking individuals to the social order. People are socialized and enculturated by others and internalize norms. Both in childhood and adulthood (Brim 1968) individuals are socialized into new roles and become good citizens, that is, socialized actors. The problem lies in the consensual nature of social organization, the conformity of actors, and how to deal with variability. At a deeper level the question that must be resolved theoretically is the relationship between norms and action. Although we cannot resolve this issue here, a number of alternatives are available largely rooted in non-functionalist paradigms in sociology and anthropology which suggest creative solutions.

Our second lesson, of applicability of the life course to all kinds of societies, is a challenge for empirical studies. At issue is that the life course has become institutionalized in Euro-American states. With our model of the life course having been worked out in the context of these societies, we should be cautious and aware of generalizability and compatibility issues as we take the model to other societies. What triggers the caution is the fact the life course has been institutionalized and its standardization is rooted in the structure of industrial states. Our first warning sign of impending institutionalization was the appearance of childhood as a life stage (Aires 1962). What came next was greater rationalization and structuring of the life course (Meyer 1986). This change is a product of the political economy of the industrial state. People are individuated as citizens, and not as members of competing collectivities such as families, communities, etc. In the transformation of the life course, chronological reference increased, it became rationalized, and individuated (Kohli 1986). States have structured the life course through child labor laws, finely graded and mandatory education, formalized rules concerning seniority and career sequencing, and formalized retirement and entitlement programs (Mayer and Muller 1986).

In the absence of the state's penetration into the lives of individuals, what happens to the indicators of social time? Where individuals are submerged in collectivities such as lineages, their biographies become secondary to the corporate needs of their economic and reproductive needs. Domestic economies do not need the integration of industrial economies. Thus we can expect less precision in social time where there are no "lock-step markers" such as formal education-careers-retirement. Although salient for non-industrial contexts, the same issues may be relevant for post-modern industrial societies. With increased divorce, later ages for child bearing, and more people remaining single, the family cycle is certainly becoming a less predictable calibrator of the post-modern social clocks.

Adequacy of representation and appropriateness of the life course model are issues which can only be empirically resolved. Since its inception in the 1960's and 1970's, the life course perspective has stimulated research which is only loosely integrated. Most of this research is empirically anchored within Euro-American industrial states. Paradigms which are functional in nature, have limitations if they are of only one social type. To be most effective, functional perspectives should be comparative (e.g., across cohorts, time, social divisions or across cultures). Indeed Radcliffe-Brown's agenda (1957) called for comparisons in the "social morphology" and "social physiology" of diverse societies to solve major theoretical problems once sufficient empirical evidence had been gathered.

For development of theory within the life course paradigm, we need comparative studies of the most diverse nature. By empirically investigating how life courses vary across cultures and within cultures we can begin to generate and evaluate hypotheses about the cultural and ecological forces which shape life courses. We can also evaluate theories linking the social, cultural and individual variation in life courses to the aging process.

## REFERENCES

Ables, R. P., L. Steel and L. L. Wise (1980) Patterns and Implications of Life-Course Organization: Studies from Project TALENT. In Life Span Development and Behavior. P. Baltes and O. G. Brim, eds. (Volume 3.) New York: Academic Press.

Aires, M. (1962) Centuries of Childhood. New York: Random House.

Allport, G. (1942) The Use of Personal Documents in Psychological Science. New York: Social Science Research Council, Bulletin 49.

Baltes, P. B. and J. R. Nesselroade (1979) History of Rationale of Longitudinal Research. In J. R. Nesselroade and P. B. Baltes, eds. Longitudinal Research in the Study of Behavior and Development. New York: Academic Press.

Baltes, P. B., and K. W. Schaie (1973) Life-Span Developmental Psychology: Personality and Socialization. New York: Academic Press.

Baxter, P. T. W. (1978) Boran Age-Sets and Generation Sets: Gada, a Puzzle or a Maze? In Age Generation and Time. P. T. W. Baxter and U. Almagor, eds. New York: St. Martin's Press.

Baxter, P. T. W., and U. Almagor (1978) Age, Generation and Time. New York: St. Martin's Press.

Benet, R. (1974) Abkhasians: The Long-Living People of the Caucasus. New York: Holt, Reinhart, Winston.

Bernardi, B. (1985) Age Class Systems: Social Institutions and Polities Based on Age. London: Cambridge University Press.

Bertaux, D. (1981) Biography and Society. Beverly Hills: Sage.

Birren, J. E., W. R. Cunningham, and K. Yamamoto (1983) Psychology of Development and Aging. Annual Review of Psychology 34:543-575.

Bohannan, P. J. (1953) Concepts of Time Among Tiv of Nigeria. Southwestern Journal of Anthropology 9:251-262.

Brim, O. G. (1968) Adult Socialization. In Socialization and Society. J. A. Clausen, ed. Boston: Little, Brown and Company.

Buchler, I. A., and Selby, H. A. (1978) Kinship and Social Organization. New York: MacMillian Company.

Buhler, C. (1935) The Curve of Live as Studied in Biographies. Journal of Applied Psychology 19:405-409.

Busse, E., and G. Maddox (1985) The Duke Longitudinal Studies of Normal Aging (1955-1980). New York: Springer Publishing Company.

Cain, L. D. (1964) Life Course and Social Structure. In Handbook of Modern Sociology. R. E. L. Faris (ed.) Chicago: Rand McNally.

Callahan, E. J., and K. A. McClusky (1983) Life Span Developmental Psychology: Nonnormative Life Events. New York: Academic Press.

Cancian. F. (1975) What are Norms? A Study of Beliefs and Action in a Maya Community. New York: Cambridge University Press.

Campbell, R. T., J. Abolafia, and G. L. Maddox (1985) Life Course Analysis in Social Gerontology: Using Replicated Social Surveys to Study Cohort Differences. In Gender and the Life Course. A. S. Rossi, ed. New York: Aldine.

Dannefer, D. (1988) What's in a Name? An Account of the Neglect of Variability in the Study of Aging. In Emergent Theories of Aging. J. E. Birren and V. L. Bengtson (eds.) New York: Springer.

Datan, N. and L. H. Ginsberg (1975) Life-Span Developmental Psychology: Normative Life Crises. New York: Academic Press.

Demos, J. and S. Boocock (1978) Turning Points: Historical and Sociological Essays on the Family. Chicago: University of Chicago Press.

Dollard, J. (1935) Criteria for Life History. New Haven: Yale University Press.

Dyk, W. (1938) Son of Old Man Hat: A Navajo Autobiography Recorded by W. Dyk. New York: Harcourt.

Eichorn, D., J. Clausen, N. Haan, M. Honzik, and P. Mussen (1981) Present and Past in Middle Life. New York: Academic Press.
Elder, G. H. (1974) Children of the Great Depression. Chicago: University of Chicago Press.
Erickson, E. H. (1959) Identity and the Life Cycle. Psychological Issues, 1: 18-164.
Evans-Prichard, E. E. (1940) The Nuer. Oxford: Clarendon.
Featherman, D. L. (1981) The Life Span Perspective in Social Science Research. New York: Social Science Research Council.
Foner, A. and D. I. Kertzer (1978) Transitions over the Life Course: Lessons from Age Set Societies. American Journal of Sociology 83:1081-1104.
Fortes, M. (1959) Oedipus and Job in West African Religion. Cambridge: Cambridge University Press.
Fry, C. L. (1976) The Ages of Adulthood: A Question of Numbers. Journal of Gerontology 31:170-177.
Fry, C. L. (1980) Cultural Dimensions of Age. In Aging Culture and Society: Comparative Perspectives and Strategies. C. L. Fry, ed. New York: Praeger.
Fry, C. L. (1983) Temporal and Status Dimensions of Life Cycles. International Journal of Aging and Human Development. 17:281-300.
Fry, C. L. (1986) The Emics of Age: Cognitive Anthropology and Age Differentiation. In New Methods for Old Age Research. C. L. Fry and J. Keith, eds. South Hadley: Bergin and Garvey Publishers.
Fry, C. L. (1988) Theories of Age and Culture. In Emergent Theories of Aging. J. E. Birren and V. L. Bengtson, eds. New York: Springer Publishing Co.
Fry, C. L., J. Keith (1982) The Life Course as a Cultural Unit. In Aging From Birth to Death. Vol II: Sociotemporal Perspectives. M. W. Riley, R. Ables, and M. S. Teitelbaum, eds. Boulder: Westview Press.
Glenn, N. D. (1976) Cohort Analyst's Futile Quest: Statistical Attempts to Separate Age, Period, and Cohort Effects. American Sociological Review 41:900-904.
Glick, P. C. (1977) Updating the Family Life-cycle. Journal of Marriage and the Family 39:5-13.
Gottschalk, L., C. Kluckhohn, and R. Angell (1945) The Use of Personal Documents in History, Anthropology, and Sociology. New York: Social Science Research Council, Bulletin 53.
Goulet, L. R., and P. B. Baltes (1970) Life-Span Developmental Psychology Research and Theory. New York: Academic Press.
Guemple, L. (1969) Human Resource Management: The Dilemma of the Aging Eskimo. Sociological Symposium. 2:59-74.
Guemple, L. (1980) Growing Old in Inuit Society. In Aging in Canada: Social Perspectives. V. W. Marshall (ed.). Don Mills, Ontario: Fitzhenry and Whiteside.
Hagestad, G. O. and B. L. Neugarten (1985) Age and the Life Course. In The Handbook of Aging and the Social Sciences. R. H. Binstock and Ethel Shanas, eds. New York: Van Nostrand Reinhold.
Hareven, T. K. (1978) Transitions: The Family and the Life Course in Historical Perspective. New York: Academic Press.
Hareven, T. K. (1982) Family Time and Industrial Time. Cambridge: Cambridge University Press.
Hart, C. W. M., A. R. Pilling, and J. C. Goodale (1988) The Tiwi of North Australia. Third Edition. New York: Hold Reinhart and Winston.
Hauser, P. M. (1976) Aging and World-Wide Population Change. In The Handbook of Aging and the Social Sciences. R. H. Binstock and E. Shanas, eds. New York: Von Nostrand Reinhold.
Hendricks, C. D. and J. Hendricks (1976) Concepts of Time and Temporal Construction among the Aged with Implications for Research. In Time Roles and Self in Old Age. J. F. Gubrium, ed. New York: Human Sciences Press.
Hendricks, J., and M. Seltzer (1986) Aging and Time. American Behavioral Scientist 29:6. (Special edited issue).
Hogan, D. P. (1981) Transitions and Social Change: The Early Lives of American Men. New York:

Academic Press.
Ikels, C., J. Keith and C. L. Fry (1988) The Use of Qualitative Methodologies in Large-Scale Cross-Cultural Research. In Qualitative Gerontology. S. Reinharz and G. D. Rowles, eds. New York: Springer Publishing.
Kardiner, A. (1945) Psychological Frontiers of Society. New York: Columbia University Press.
Keith, J. (1982) Old People as People: Social and Cultural Influences on Aging and Old Age. Boston: Little Brown.
Keith, J. and C. L. Fry (1988) Mobility, Stability and Strategies for Well-Being in Old Age: Evidence from Cross-Cultural Research. Presented at at the Gerontological Society of America.
Keith, J., C. L. Fry and C. Ikels (1990) Community as Context for Successful Aging. In The Cultural Context of Aging J. Sokolovsky, ed. South Hadley: Bergin and Garvey Publishers.
Kertzer, D. I. and J. Keith (1984) Age and Anthropological Theory. Ithaca: Cornell University Press.
Kertzer, D. I. and O. B. B. Madison (1981) Women's Age Set Systems in Africa. In Dimensions, Aging, Culture and Health. C.L. Fry, ed. New York: Praeger
Kohli, M. (1986) The World We Forgot: A Historical Review of the Life Course. In Later Life: The Social Psychology of Aging. V. W. Marshall, ed. Beverly Hills: Sage.
LaFarge, O. (1929) Laughing Boy. Boston: Houghton Mifflin.
Langness, L. L. (1965) The Life History in Anthropological Science. New York: Holt.
Langness, L. L. (1981) Lives: An Anthropological Approach to Biography. Novato, CA: Chandler and Sharp.
Leach, E. R. (1961) Rethinking Anthropology. New York: Humanities Press.
LeVine, S. and R. A. LeVine (1985) Age Gender and the Demographic Transition: The Life Course in Agrarian Societies. In A. S. Rossi, ed. Gender and the Life Course. New York: Aldine.
Maddox, G. L. and R. T. Campbell (1985) Scope, Concepts, and Methods in the Study of Aging. In Handbook of Aging and the Social Sciences. R. H. Binstock and E. Shanas, eds. New York: Von Nostrand Reinhold.
Maxwell, R. J. (1972) Anthropological Perspectives. In The Future of Time. H. Yaker ed. Garden City N.J.: Anchor Books.
Mayer, K. U. and W. Muller (1986) The State and the Structure of the Life Course. In Human Development and the Life Course: Multidisciplinary Perspectives. A. B. Soresen, F. W. Weinert, and L. R. Sherrod, eds. Hillsdale, N.J.: Lawrence Erlbaum Associates.
Meyer, J. W. (1986) The Self and the Life Course: Institutionalization and its Effects. In Human Development and the Life Course: Multidisciplinary Perspectives. A. B. Soresen, F. W. Weinert, and L. R. Sherrod, eds. Hillsdale, N.J.: Lawrence Erlbaum Associates.
Moore, A. (1973) Life Cycles in Atchalan: The Diverse Careers of Certain Guatemalans. New York: Teachers College, Columbia University Press.
Neugarten, B. L. (1968) Middle Age and Aging: A Reader in Social Psychology. Chicago: Chicago University Press.
Neugarten, B. L. (1982) Age or Need? Public Policies for Older People. Beverly Hills: Sage.
Neugarten, B. L. and N. Datan (1973) Sociological Perspectives on the Life Cycle. In Life-Span Developmental Psychology: Personality and Socialization. P. B. Baltes and K. W. Schaie, eds. New York: Academic Press.
Neugarten, B. L., and J. W. Moore (1968) The Changing Age Status System. In Middle Age and Aging. B. L. Neugarten, ed. Chicago: University of Chicago Press.
Neugarten, B. L., J. W. Moore and J. C. Lowie (1965) Age Norms, Age Constraints, and Adult Socialization. American Journal of Sociology 70:710-17.
Neugarten, B. L. and W. A. Peterson (1957) A Study of the American Age Grading System. In Proceedings of the 4th Congress of the International Association of Gerontology, Vol 3.
Nydegger, C. (1981) On Being Caught Up in Time. Human Development 24:1-12.
Nydegger, C. (1986a) Age and Life Course Transitions. In New Methods for Old Age Research. C. L. Fry and J. Keith,eds. South Hadley: Bergin and Garvey Publishers.

Nydegger, C. (1986b) Timetables and Implicit Theory. American Behavioral Scientist 29:1986.
Ortner, S. B. (1984) Theory in Anthropology Since the Sixties. Studies in Comparative Societies and History. 26:126-166.
Passuth, P. M. and V. L. Bengtson (1988) Sociological Theories of Aging: Current Perspectives and Future Directions. In. J. E. Birren and V. L. Bengtson, eds. Emergent Theories of Aging. New York: Springer Publishing.
Plath, D. and K. Ikeda (1975) After Coming of Age: Adult Awareness of Age Norms. In Socialization and Communication in Primary Groups. T. R. Williams, ed. Mouton: The Hague.
Radcliffe-Brown, A. R. (1929) Age Organization Terminology. Man 29:21.
Radcliffe-Brown, A. R. (1957) A Natural Science of Society. Glencoe: The Free Press.
Radin, P. (1920) The Autobiography of a Winnebago Indian. University of California Publications in American Archaeology and Ethnology. vol 16: 381-473.
Radin, P. (1926) Crashing Thunder, The Autobiography of an American Indian. New York: Appleton.
Reinert, G. (1979) Prolegomena to a History of Life-Span Developmental Psychology. In Life-Span Development and Behavior. (Volume 2). P. B. Baltes and O. G. Brim, Jr., eds. New York: Academic Press.
Riley, M. W. (1979) Introduction. In Aging From Birth to Death: Interdisciplinary Perspectives. M. W. Riley (ed.) Boulder: Westview Press.
Riley, M. W. (1986) Age Strata in Social Systems. In Handbook of Aging and the Social Sciences. R. H. Binstock and E. Shanas, eds. New York: Von Nostrand Reinhold.
Riley, M. W., M. E. Johnson and A. Foner (1972) Aging and Society: A Sociology of Age Stratification. (Volume 3). New York: Russell Sage.
Rodgers, W. L. (1982) Estimable Functions of Age, Period, and Cohort Effects. American Sociological Review 47:774-787.
Rosow, I. (1978) What is a Cohort and Why? Human Development 21:65-75.
Schaie, K. W. (1965) A General Model of the Study of Developmental Problems. Psychological Bulletin 64:92-107.
Schneider. D. M. (1965) Some Muddles in the Models: Or How the System Works. In The Relevance of Models for Social Anthropology. M. Banton, ed. New York: Praeger.
Shock, N. W., R. C. Greulich, and R. Andres (1984) Normal Human Aging: The Baltimore Study of Aging. N.I.H. Publication No. 84-2450. Washington DC: U. S. Government Printing Office.
Simmons, L. W. (1942) Sun Chief: The Autobiography of a Hopi Indian. New Haven: Yale University Press.
Shostack, M. (1983) Nisa: The Life and Words of a !Kung Woman. New York: Vintage.
Smith, R. J. (1961) Cultural Differences in the Life Cycle and the Concept of Time. In Aging and Leisure. R. Kleemeier, ed. New York: Oxford University Press.
Stewart, F. H. (1977) Fundamentals of Age Group Systems. New York: Academic Press.
Thomas, W. I. and F. Znaniecki (1918-1920) The Polish Peasant in Europe and America. (Volumes 1-5). New York: Knopf.
Uhlenberg, P. (1974) Cohort Variation in Family Life Cycle Experiences of United States Females. Journal of Marriage and the Family 36:284-292.
Whorf, B. L. (1964) An American Indian Model of the Universe. In Language, Thought and Reality: Selected Writings of Benjamin Lee Whorf. J. B. Carrol, ed. Cambridge: M.I.T. Press.

# SECTION THREE

AREAL STUDIES

CRISTIE KIEFER

## 6. AGING AND THE ELDERLY IN JAPAN

It was inevitable that a close relationship should develop between American social gerontology and the ethnology of Japan. Among the many customs of the country that appeared exotic to American eyes, the veneration of the elderly attracted early attention from journalists and anthropologists alike. Around the turn of the present century, Japan's foremost literary apologist in the West, Lafcadio Hearn, was writing about the power and prestige of the elderly there (Hearn 1920), and the first national character study by an anthropologist (Benedict 1946) continued to develop the picture Hearn had offered.

The obvious potential of Japan as a subject of comparison with the West, however, posed a certain danger to scholarship. Early writings were of uneven quality, and tended to dwell on the exotic surface features of Japanese life, at the expense of detailed inquiry into the underlying structures of the culture. Although Hearn's fluency in the language and culture was impressive for example, he was more interested in telling a story than in careful comparative research on aging or any other subject. He did not seek to develop the functional significance of Japanese customs by carefully locating them in their cultural, ecological, and biological contexts, as the serious student of culture must. Other competent writers, pursuing other subjects, mentioned age related behavior in passing, and gerontologists were tempted to draw inferences from compilations of these anecdotal materials.

In the early 1960's, when cross-cultural materials entered the mainstream of social gerontology (Kleemeier 1961; Clark and Anderson 1967), the image of Japan as a gerontocratic society was already present (Smith 1961). This image had an important effect on the field. In 1968, Margaret Clark began what is probably the first controlled cross-cultural fieldwork on aging. She wanted to look at intergenerational relations in three cultures, stratifying for achievement motivation and veneration of the aged. She wanted to know, among other things, whether American-style competitiveness and material acquisitiveness interfered with the status of the elderly. She chose Japanese Americans as her high-achievement, high-veneration sample, and I joined her to do the fieldwork on that sample (Kiefer 1974).

About the same time, Cowgill and Holmes (1972) published the first systematic statement of a theory that had been emerging in social gerontology, now called Modernization Theory. They held that industrialization necessarily undercuts the status of the aged by nuclearizing the family, prolonging life, and placing a premium on social and technological innovation, productivity, and wealth. Their groundbreaking work drew support from a study by David Plath of Japanese popular culture. This was an unexpected source of support, as it seemed to contradict the accepted picture of Japan as an exception to the industrial rule. With his unusual bilingual and biliterate skills, Plath had analyzed Japanese folklore, newspapers and magazines, films, and fiction to show that attitudes toward the aging were at best highly ambivalent there (Plath 1972).

Duke sociologist Erdman Palmore, however, who had spent his early years in Japan as the son of Christian missionaries, quickly challenged the Cowgill and Holmes thesis in general, and Plath in particular, from the vantage point of Japanese ethnography and vital statistics (Palmore 1975).

I will examine the controversy over the status of Japan's aged later on, when I discuss security, prestige, and power. At this point, there is still much work to be done to resolve it. The early literature on the topic was built upon repetitive use of a narrow base of Western writing. The Japanese government began financing studies of attitudes of, and toward, the aged in the postwar era, supplementing statistical analyses of income, housing, and other indirect measures of prestige – studies of the kind that have long been available on all age groups – but the exact relevance of the attitudinal measures to the status issue is unclear. Descriptive studies of the elderly and their environments would clarify the issue, but such studies by social scientists have only begun to emerge in the 1980's, and are still quite sketchy. There are some excellent literary descriptions of the social relations of old people by Japanese authors, and some of these are available in English translation. It would be especially worthwhile for gerontologists to compare systematically Inouye's *Chronicle of My Mother* (1982), Ariyoshi's *Man in Ecstasy* (1972), and Niwa's *The Hateful Age* (1962).

Good written and filmed ethnographic studies of Japanese life are abundant, reaching back to John Embree's work in the 1930s (Embree 1939), and covering urban, suburban and rural neighborhoods, factories, schools, apartment buildings, professions, and individual and family biographies. I will not review or list these studies here, but only note that many of them contain glimpses of aging and the aged that help to round-out pictures that emerge from more strictly gerontological studies. For example, Plath's biographies of middle aged Japanese (1980) and Lebra's study of modern women (1984) provide diverse but equally useful views of the interactions between life cycle and history.

A panoramic view of anthropology-and-gerontology in Japan to date, then, reveals two broad periods. First, a period from the early 1960s to the early 1970s, of anecdotal reporting by Western journalists and anthropologists on customs and language related to aging. This material had a significant impact on modernization theory. Second, a period of research, from the early 1970s to the present, by Japanese and American scholars on attitudes and material conditions related to aging. The Japanese gerontologists tend to focus on health and welfare problems created by the aging of the Japanese population (which I shall discuss under Dependency, below), and to leave descriptive accounts of aging to the novelists and journalists. Most of this material is published in Japanese. The American researchers are more interested in the relevance of the data to theoretical discussions of status and prestige. A third period of scholarship may be newly under way – a period of descriptive specifically gerontological research by American-trained scholars, beginning in the early 1980s. In the sections that follow, I shall summarize the findings of this research, but I must begin by sketching its historical and cultural context.

### THE CULTURAL BACKGROUND

Gestures, behaviors and attitudes, and the material arrangements of human architecture, travel, and communication are always situated in complex systems of expectation and interpretation and draw their significance from those systems. Although this scarcely needs to be said to anthropologists, it is not always easy for others to grasp. Perhaps it can be understood best if the relationship between behaviors (such as addressing an old person with an honorific title) and their cultural contexts are compared to the relationship between words and their sentences. An individual word usually covers a broad, vague range of meaning, which overlaps with the meanings of some other words, but is unique in some of its potential meanings. When we use a word in a sentence, we create a context for it which eliminates many of its potential meanings, and often gives it a very specific semantic content. By itself, a word usually evokes those meanings we give it most frequently in our speech, so that if I say, "Think of red," for example, my listener probably thinks of a very common shade, like the red of the American flag. But if I say, "Her hair was red," or "He was riding a red roan stallion," my listener will think of very different colors. If I say, "I really saw red when she accused me of that," or "The Reds have rigged the election," he may not think of any color at all.

In just this way, behaviors like bowing, or giving someone a birthday party, may occur in different situations, with different meanings. If they occur in an unfamiliar culture, we might recognize their form but completely miss their meaning.

This is precisely the problem we confront when we look at isolated age-related behaviors in Japanese culture. Because of the complexity of cultural systems, and the plasticity of meanings within a culture, even a detailed ethnography cannot completely eliminate this problem. Here I only want to try to convey a sense of the problem by sketching some Japanese cultural traditions of particular importance in the interpretation of age-related behavior.

Much has been written about the group-centeredness and hierarchical organization of Japanese society. The work of Chie Nakane (1972) gives an excellent summary. These features of social organization are pervasive, historically deep, and largely unconscious, and they closely affect nearly all social behavior. They are in some ways superficially similar to features of preindustrial European societies, and their similarity is often taken for functional equivalence, as when they are described as "preindustrial" or simply "traditional." However, the *Confucian base* and the *corporate emphasis* of Japanese ethics make these patterns somewhat unique there.

This uniqueness is strengthened, moreover, by the relative homogeneity of Japanese culture. A small island nation, Japan was centrally ruled by a powerful and rigid bureaucracy, and largely cut-off from contact with the outside world, from 1603 until 1868. The result was a language, a set of customs, and a view of the world that is remarkably uniform throughout the archipelago. As a result, Japanese behavior is more often based on accurate perceptions of uniform unstated social expectations than is the case in most Western cultures.

*Confucianism*

Confucianism was borrowed from China in the fifth century, along with writing, Buddhism, and many other features of historic Japanese culture. Confucian values emphasize the proper performance of well defined social roles at all levels of society, from the noblest to the humblest. The most important of all relationships is that of parent and child; and the norms call for respect, obedience, support, and kindness toward parents regardless of the ages of the generations in question.

This system of ethics was embedded in, and in turn supported, an agrarian village organization in which all families were integrated into large, landed kin units. Authority was regulated by age-grading, and by a kinship system based on the patrilineal extended family, or *ie*. The *ie* was a strong hierarchy of interrelated households, within which seniority, primogeniture, and male dominance characterized formal authority.

Like many societies with a strong descent-based social organization, the Japanese also maintained age grades. Although the details of the age-grading system differed according to locale, in a typical configuration the rank of "elder" was reached by males at the age of sixty, whereupon they assumed ceremonial roles in village life, especially religious life (Norbeck 1953). An increased freedom was granted from the constraints of younger roles, so that old people were expected to enjoy leisure, strong drink, and bawdy humor. This license was signaled by the ceremonial wearing of red clothing, a symbol of childhood. Since the dominant residence pattern was patrilocal, women entered at a somewhat younger age the powerful domestic status of mother-in-law, and commanded the duties of their sons' wives.

The rights and duties of the elders were often explicit and formal, and to some extent they assured the social integration into wider society of at least those elders who had fulfilled the normal familial roles appropriate to their younger years. Individuals who had led exemplary lives were able to retain their prestige and power in late life through the performance of age-grade functions. The role of the artist, scholar, priest, or politician, for example, still tends to confer increasing prestige as the individual ages. The Japanese government's naming of very old artists as "living cultural treasures" formalizes this custom (Smith 1961).

An important feature of this social organization which distinguishes it from traditional Western societies was its pervasiveness. Whereas inheritance patterns gave some authority to the elders of landed families in premodern Europe, many peasants did not control land, and this fact limited the power of the elders. In Japan, by contrast, the integration of the lower peasantry into the landholding system reinforced the confucian emphasis on universal respect for the elders.

This set of traditions has produced modern results in custom if not in law. It has resulted in a "deep structure" of ethics in which the proper performance of social roles is essential for social success *in all classes*, and in which the roles of parent and child are salient throughout life. This has buffered, but certainly not eliminated, the tendency of economic values to supersede family values in industrial society. From the viewpoint of law, the confucian tradition together with the feudal origins of the modern Japanese state, resulted in the granting of extensive rights to the elderly, as I shall discuss shortly.

*Corporate Emphasis*

Confucianism and the traditional social organization thereby levied heavy requirements of respect and care for the old upon the young; but the old themselves were by no means without obligations under this system. Another traditional principal that guided social life was the subordination of the individual, regardless of formal office or authority, to the well-being of the group. Although households were theoretically ruled by the oldest active male, for example, decisions were often made for the household by the most competent member. A person in authority who wielded that authority unfeelingly or neglected the well-being of subordinates, came in for general social disapproval, and sometimes punishment at the hands of the community. Even today, this value regulates not only families but many other types of groups, including businesses, voluntary organizations, and communities. Consensus is considered the ideal form of group decision making. This feature of social organization, called "corporate emphasis" by Befu (1962), separates power and prestige to some extent. The older person who fails in his or her responsibilities is not exempt from ridicule and even rebellion. As we shall see, this cultural complex is expressed today in the strong sense of obligation which older people bring to their work and family roles.

In America, where individual social behavior is typically the result of personal selection from a wide range of acceptable options, we tend to assume that personal feelings precede, and therefore support, interpersonal behaviors. The correctness of this assumption in our own culture is often questionable, and it is even less valid when applied to Japan. There, a massive weight of widely shared expectation more often takes precedence over conscious personal choice.

This group-centeredness and homogeneity of values lends a certain stability to Japanese society – social rules change slowly. But they do change. An understanding of the modern status of the elderly requires some discussion of history.

### HISTORICAL BACKGROUND

Today Japan has a mature industrial economy, enjoys a stable, popular government, and has a living standard, an environment, and a per capita income that are among the best to be found anywhere in the industrial world. Its unique features as a modern society are its high population density, its language and art, and those features of interpersonal relations that I have just discussed. Two features of modern Japanese history must be mentioned, however, before we can discuss the condition of the Japanese aged: (a) the recency of industrialization, and (b) the underwriting of confucian values in modern law.

In 1868, when the feudal warrior class was displaced from power, Japan had already developed many features of a capitalist society: a money economy, national markets, credit buying, large ramified commercial corporations, extensive urban life, and specialized cash farming, for example. Still, many features of the industrial economies of the Western nations had yet to appear. There was little foreign trade, little practical education, few political freedoms, poor nutrition and public health, and a very rudimentary and labor-intensive technology. Some classes, such as the successful urban merchants and upper samurai, were quite well off. But the great mass of Japanese people were nonliterate peasants and laborers with a very low standard of living. For all classes infant mortality was high, and life expectancy at birth was short.

By the early 1900s, Japan had achieved European style military strength, and had developed an export trade that threatened Western competitors. Necessarily it had in the process become a political rival of the industrial nations. This transformation took place too fast to be a deep social evolution. Rather, it was the grafting-on of a technology to the preindustrial social organization and value system.

Respect for traditional forms of social relationship remained strong: master and disciple, patron and client, priest and parishioner, trader and partner, neighbor and neighbor, continued many ancient forms of exchange and expressions of mutual respect. This offered a strong contrast with the urban West, where free markets and economic mobility had already led to more strictly monetary and bureaucratic forms of exchange. Even today Japanese social relationships show a kind of civility that, while cumbersome, gives them greater durability than the transactions of Western technocracies.

The speed of this transformation in some ways gave Japan an advantage over other industrial countries. Starting late, the Japanese were able to copy economic and political systems that had already achieved workable procedures through trial and error. The Japanese leaders were careful to choose national financial policies and civil laws that preserved traditions, kept social life stable, and protected the most vulnerable populations. There were periods of civil unrest, but less than there had been in the Western cultures during the period of industrialization.

One of the mechanisms consciously used by the government to preserve social order was the propagation of traditional values, both through education and through the legal codification of the rights and obligations of family members. The values of family solidarity and Confucian filial piety were drawn upon to modulate the disruptive nature of social change. In 1871 the system of family registers was established, requiring the details of residence, employment, and marriage of every citizen to be registered in the *koseki*, or family register. It also gave the household head power to expel persons from his residence. In 1899, the civil code was revised to require families to care for their aged members. In this new law, the parents of household heads had legal priority over wives and children, and parents-in-law over brothers and sisters (Kinoshita 1984).

Although the aged actually retain few such legal rights today under the postwar constitution, the effect of these policies was to lend a sense of jural entitlement to the already respected position of the aged in Japanese society. As we have seen, traditionally, household heads carried responsibility for the wellbeing of household members, and could not be neglectful or cruel to wives, parents, siblings, servants or children without risking strong social censure. The laws gave an added measure of responsibility to kin, placing the family as a safety net under the destitute – at least that great majority of the destitute who escaped the terrible misfortune of being without a family

## MODERN TIMES

The accelerated industrialization and modernization of Japan produced a compressed population explosion. The total population of 30 million in 1860 more than tripled in the next 100 years, and today it has quadrupled to 122 million. Since World War II, slowed birth rates and greatly improved public health and medicine have shifted the growth from the lower end to the upper end of the lifespan, so that the over-65 population has increased faster than other segments. Today there are about thirteen million people in this segment, representing 10.3% of the population, up from 4.9% in 1950. In twenty years, the proportion of elderly is expected to reach 20%, twenty-seven million people, and two million more than the under-14 age group (Ministry of Health and Welfare 1987: 20).

The speed of this graying trend has been one of the most challenging aspects of Japan's phenomenal postwar development. The well-being of the elderly has become a constant preoccupation of the nation.

*Security, Integration, Power, and Status*

Most anthropological studies of aging in the 1960s and 1970s centered attention on the conditions that promote or detract from the well-being of the elderly. As I have mentioned, American studies of Japan were no exception, but dealt mainly with the relevance of the Japanese case to modernization theory. Elsewhere (Kiefer 1990) I have discussed the history and problems of modernization theory in general, pointing out that the complex dependent variable – the well-being of the elderly – can be clarified by discussing some of its components separately. Keeping the general framework of modernization theory in mind, then, let us review contemporary knowledge of Japan's aged from four perspectives: security, integration, power, and prestige.

*Security*

All old people share the insecurity of having used up more of the finite human lifespan than their younger contemporaries. But the elderly differ among themselves in their access to the vital resources that are needed to reach their potential of health and longevity, as well as the resources needed to assure that the time remaining to them shall be endurable – that they shall have what I call *value security*. Paradoxically, this inequality is often heightened by modern medicine, which can add years of fragile, poor quality life to an existence that would have been shorter under natural conditions. In this sense, the insecurity of the Japanese elderly is high. Life expectancy at birth is 74.8 years for men and 80.5 years for women, making Japan the world's second most longevous nation (Ministry of Health and Welfare 1987). About 4% of those over 65 are in hospitals, and about 1.8% in other institutions at any given time, giving a grand total of three quarters of a million institutionalized elderly.

Of course the majority of those over 65 are healthy. These healthy people rarely express worries about dying, but they often say that their greatest worry is that their lives will become meaningless long before they die (Kiefer 1987). Certain Buddhist temples and Shinto shrines have come to be associated with the granting of speedy death, and these are visited by a steady stream of old people.

Such behavior simply reflects the keen sense of obligation the elders feel not to become a burden on their children, not necessarily a weariness with life. Suicide rates for Japan as a whole are high, running at about 13 per 100,000 for all age groups, and they rise in youth and old age. About twenty of every 100,000 women between 60 and 64, for example, take their own lives; but this cannot be taken as a sign of depression or alienation. As DeVos and Wagatsuma (1973) have argued, Japanese suicide often indicates an *over*involvement in social roles.

Taking all these circumstances together, it seems unfeeling of the gerontologist to measure security in terms of material resources alone. Later I will return to the question of dependency. I believe the value security of the Japanese aged is improving as better ways are being found to keep the frail elderly at home and in control of their lives.

Of course material resources are important as well. I have discussed the norms giving responsibility for the care of the aged to their children, and these norms contribute substantially to a sense of material security in old age. However, this situation is of little comfort to those elderly who are alienated from their children, or who have none, or whose children are themselves poor. This contributes to the fact that one out of every three people over sixty-five works. Most pensions in Japan are still well below American levels. Okamura (1987) found that 34.5% of a sample of 501 retirees from companies with fixed-age retirement rules had no pensions at all, perhaps reflecting the typical retirement age of 55, and the "gap" of 5 years between this age and the beginning of government pensions. Of workers who retire at 55, about half go into second jobs (Kii 1979), and of these recycled workers, the majority work at less prestigious jobs than before, with less pay (Naoi 1987). Hashimoto (1986) estimates that 42% of Japan's elderly receive welfare payments, indicating that their personal income is less than $2,900 per month.

High rates of employment should not be taken as simple evidence of economic need, however. About the same proportion of Americans (42.1%) and Japanese (44.1%) over the age of sixty say they need to work for money, but 38% of elderly Japanese who work say they do so to preserve their health, whereas only 14% of older American workers say this (Prime Minister's Office 1982). This may reflect the greater responsibility Japanese elderly feel in their family and community roles.

*Integration*

Studies of the social integration of the elderly form an important component of anthropological gerontology. It is generally assumed, although largely unproven, that old people are happier when they are integrated into families and communities. It is also generally assumed that the degree of integration can be measured in a number of ways. Housing arrangements, frequency of contact with younger people, employment, and group activities are common kinds of measures.

Most students of Japan portray the elderly as well integrated into their society. In 1985, 61.8% of those over 60 lived with children nationwide (Ministry of Health and Welfare 1987: 27). Although this is a noticeable decline from the 66.7% reported in 1980, it is still very high compared with western countries, and especially with the United States, where the rate is about 14%. Another indication of relative integratedness is the employment rate of the

Japanese elderly, mentioned earlier.

However, the picture of integration is not uniform. A number of authors (Sparks 1975; Kii 1979) have pointed out the loss of status, income, and social interaction that accompanies forced retirement for men, usually at age 55. Moreover, in spite of the popularity of old people's clubs in Japan (Maeda 1975), elderly Americans are about six times as likely to be participating in social activities outside the home (Prime Minister's Office 1982: 135-137). A comparative study of elderly in West Haven, Connecticut and Odawara, Kanagawa Prefecture (Hashimoto 1986) found that the Americans had far larger and closer kin and friendship networks than the Japanese.

Even the rate of two-and three-generation households may be misleading. Co-residence with children for the Japanese elderly is more likely to be widowed. Those who live with their offspring are older and sicker than those who do not (Koyono *et al.* 1986), indicating that co-residence is less a matter of mutual preference than a matter of medical necessity. This notion is borne out by Ikegami (1982: 2005), who found that elderly were more likely to be hospitalized if they were living alone.

Thus there are factors other than affection that determine housing patterns, including the relative availability of health care and housing choices. The high cost of land and the relative absence of services for independent living of the frail elderly undoubtedly contribute to the picture of integratedness painted by household structure. Kinoshita (1984) found that many old people living in a retirement community had moved there out of fear of their impending physical frailty, and that many of them felt lonely and socially ill at ease in this setting.

A problem that cross-cultural gerontology has yet to address, to my knowledge, is that of measuring what is possibly the most important ingredient of social integration, the quality of social contacts. The physical separation of families can mean *improved* integration, if it leads to improved rapport between generations. "Living separately" can mean living next door (probably too close), across the country (too far), or across town (about right). When Imamura (1987: 83) asked urban house-wives where they wanted to live in old age, almost 60% said they preferred living separately from their children.

The ideal arrangement in Japan, as in America, seems to be for the two generations to live "close but separate." This, too, is probably less difficult in America. Most working-age Japanese live in densely packed metropolitan areas where independent housing is beyond the means of the typical retiree, even if he or she should want to move closer to children. This situation is reflected in higher rates of independent living in rural areas (Maeda 1983; Ikegami 1982).

Even when propinquity to children is an economic option, it often puts older women in a dilemma for two reasons: First, they would have to leave neighborhood friends, who might actually be emotionally closer than their children (Imamura 1987). Second, custom suggests that they should put priority on ties with their eldest son's family; but the absence of the son from the household most of the time would actually bring them into close contact with daughters-in-law, and perhaps away from daughters, with whom Japanese women tend to feel a natural closeness. A shift from gerontocratic values in the family to more democratic ones in recent years seems to have taken much of the fun out of being a mother-in-law. Men probably feel this dilemma less acutely, for several reasons: They are less likely to have close friends in their long-term neighborhood, they are expected to be less involved in the affairs of the household, and of course they are more likely to have a wife to look after them regardless of where they live.

Other considerations such as money, housing quality, transportation and communication options, health, common interests, and free time will play into such equations. A Japanese government survey made an intriguing attempt to measure the quality of intergenerational relations by asking old people the question, "Do you think your children are very interested, somewhat interested, not very interested, or not interested at all in your well-being?" Eighty-four percent of an American sample, versus sixty-eight percent of the Japanese, chose "very interested" (Prime Minister's Office 1982: 210). The difference may simply reflect the relative accessibility of life style options for the two generations in America, and their consequent ability to find a mutually rewarding mixture of closeness and distance.

Of course there are many dimensions to interpersonal relationships aside from their strength and comfort. Role expectations can be rigid and heavy, and can make a burden even of relations between people who like and understand each other. The competence of individuals to help one another varies according to ability, resources, and competing demands.

To get some sense of these dimensions of integration in Japan, one can learn from such descriptive accounts of family life as Inouye's fine grained and apparently factual essays (1982). In following the senile decline of the novelist's mother, from her seventy-fifth year to her death at eighty-nine, we see a pattern of relationships quite unlike what we would expect in an American family with a similar problem.

I suspect that my reaction to Inouye's *Chronicle of My Mother* is highly individual as well as American, but I can convey some of what strikes me as Japanese about the "social integration" of this dependent old person with a few observations:

1. Coping with Granny is clearly the task – and a high-priority task at that – of a large group of people. The middle-aged children, their spouses, their children, the household servants, and even an estranged brother of Granny who has moved back from America, show an energetic concern for each other, as well as for the family as a whole. Granny herself, often too confused to know what the fuss is about, even seems secondary at times to their attempts to help each other. Corporate emphasis is there, all right.

2. No attempt is made to "treat" Granny's problem, or even to dia-

gnose it. Rather, the suffering and sadness of the senile woman and her harried caretakers alike seems to be accepted as natural and inevitable. Accordingly, there is an effort toward emphatic connection, toward interpreting and joining Granny's hallucinatory world. To this end, the culture provides a rich vocabulary and a full storehouse of shared images and symbols for the communication of feelings. (However, Granny is unable because of her senility to use the cultural gifts. The effect of the family's joint effort at empathy is hardly ever to draw Granny closer to the others; rather, through it they seem to draw closer to each other, the way the audience to a performance develops a collective consciousness of its own.)

3. For all their patient concern, the family accord Granny almost no rights whatsoever. She is completely excluded from all decisions, her incompetence is discussed in her presence, and there is no self-consciousness among the family about their constant controlling stratagems and ruses. The good of the family seems to be above question as a goal. True, this may be the only way they can keep their sanity under the circumstances, but it is sharply tragic when Granny ceases to recognize the rest of them as her relatives, and must nevertheless submit to their seamless collective will.

4. The force of custom is great. The family take excruciating pains to do the "right thing" socially, knowing full well Granny will make a mess of it. They seem more upset by her violations of good form – like forgetting that Grandpa ever existed – than by her alarming disappearances and exhausting nightly rounds.

Studies like this temper our attempts to compare "levels" of social integration. The satisfactoriness of life is somehow related to the availability of supportive human contact; customs clearly cause measurements of contact (source, frequency, duration) to differ from one culture to another. But just as clearly, all these contacts occur within matrices of expectation and desire that render their effect on outcomes problematic – matrices, moreover, that are sensitive to time and circumstance.

*Power and Prestige*

Social integration and security are largely outcomes of interactions within the circle of one's kin and close others. These interactions are based partly on individual resources and personal feelings, but they are also based on general cultural perceptions of worth and power related to the statuses occupied by the elderly. Let us now consider some indications of these cultural perceptions. How do Japanese view the aged *as a group*, and how does that affect their access to resources?

The Confucian tradition rewards the display of public respect toward the aged, and one rarely sees anything else in impersonal situations. Much has been made of the custom of Honor-the-Aged Day (*Keiro no Hi*) (see Palmore 1975; Palmore and Maeda 1985) and similar national customs indicating this. Among kin and close acquaintances, less formal rules apply, however. There is no scarcity of derogatory epithets like *umeboshi-baba* (dried-plum crone) or *ojin* (old codger), *yakamashii baasan* (noisy old woman) or *oban kusai* (*jiji kusai* for men), usually said of a young person who has "old fool" habits of stodginess or grumbling. Subtler and commoner insults involve addressing old people with inappropriately familiar titles, like *chan* instead of *san* or *sama*.

Valid measures of cultural perceptions are difficult to get, but suggestive findings do not show unmixed admiration for the elderly as a group. Sussman, Romeis and Maeda (1980) gave Palmore's Facts on Aging Quiz to comparable groups of Americans and Japanese, and found that the Japanese appear less knowledgeable about the aged, and more afflicted with negative stereotypes about them. Their explanation for this finding was that there is growing conflict between generations as a result of modernization influences on family economics.

While this may very well be true, I doubt that it explains the finding. Rather, Japan is a society dominated by face-to-face relationships, and the average person does not have a well developed sense of responsibility to strangers (as anyone who has tried driving a car in Japan will testify). Thus few Japanese have sympathy for the slightly battered grandmother emerging from a crowded subway during rush hour: "She should travel when there is less traffic" (see Smith 1962). Likewise, when the announcement of free medical care for the aged began to produce unacceptably high national health care costs (see *Dependency*, below), many were quick to blame the elderly themselves. In fact, the concept of full socialized health care for the elderly was not in the end supported by the medical professions, the unions, many local governments, or the Socialist Party who had proposed the idea in the first place (Campbell 1984). One also notes the relative weakness of pension laws in Japan, mentioned earlier.

Are the Japanese elderly more respected than their American contemporaries? Yes, if one is talking about socially visible behavior, including their legal rights and the control they exert within the family through the mechanism of public face. If one is talking about private beliefs and attitudes, or about the overall ability of the aged to influence their own fate, I would not like to hazard a guess.

*Dependency*

The speed with which the Japanese population has aged has resulted in a genuine crisis in health and welfare services to the old. Of the $75 billion the Japanese spent on health care in 1984, $18 billion was spent on seven million patients over seventy years of age. Thus, nearly a quarter of the nation's health cost was going to care for an age group comprising less than 10% of the national population. Moreover, the national cost of caring for the aged had increased eightfold from $2.2 billion in 1973, when free medical care for the elderly was established (Tozawa 1986a).

Part of the problem has been the lack of long term care facilities for the disabled and incurably ill elderly. In 1975, Japan had less than 42,000 skilled nursing beds, and in 1985 the number was still only 112,000 – less than 1 per hundred elderly. (Miyajima 1986) At the same time, the proportion of hospitalized elderly has remained stable at about four per hundred. In 1984 this was 820,000 old people in acute and geriatric hospitals, with an average length of stay of about 88 days. More than half of the hospitalized elderly stay longer than six months (Tozawa 1986b).

Most of the disabled elderly are cared-for at home. It is difficult to assess their numbers, because rating systems for degrees of disability vary from source to source. Maeda (1983) estimates that there were about 438,000 bedfast elderly in Japan in 1980, and that about three quarters of these were being cared-for at home. Ikegami found in one rural area that 8% of the elderly population were institutionalized, and that another 8% could be viewed as needing institutional care. If this reflects the national situation, it would mean about 1.2 million seriously ill and disabled elderly.

Elsewhere I have described in some detail the health care system for the elderly in Japan (Kiefer 1987). Both the national government and the local health districts have been experimenting with a wide variety of care systems and payment schemes for the past decade, but the problem continues to strain the capacity of a country even as successful as this one. Attempts are being made to keep old people out of institutions, through a three-pronged strategy of preventive care, rehabilitation, and extra-institutional (home care and outpatient) services. Prevention is being pursued through free health instruction, blood pressure and cancer screenings for those over 40, So far, participation rates in these plans have been disappointing (Maeda 1983).

Deinstitutionalization is being pursued through the development of day service centers, home help programs, and respite beds in nursing homes, for bedfast elderly normally cared-for at home. The day service centers are similar to American adult day health care facilities, where impaired elderly can be brought for meals, medication supervision, and other services. Health care workers from these centers also visit the bedfast elderly at home to give baths, help with meals, and other services which the family or spouse cannot provide. The number of such centers nationwide has increased more than tenfold in the last seven years, from 20 in 1979 to 210 in 1986. In the same period, the number of respite beds has increased from 5,840 to 37,346; the number of home helpers from 13,120 to 23,555. These services are administered through the local health districts (Miyajima 1986).

Rehabilitation is an obvious way of reducing the burden of caring for the disabled elderly, but Japan has been slow to use this strategy. In 1977, only 7,251 old people in Japan were receiving rehab services (Prime Minister's Office 1980). I have argued elsewhere (Kiefer 1987) that there may be cultural reasons that interfere with the idea of rehabilitation; however, progress is being made. The number of rehabilitation centers in the local health districts is rapidly growing (from 798 in 1983 to 2,451 in 1986). Miyajima (1986) discusses a model geriatric care system which would add a new type of institution, the Geriatric Health Facility (*rojin hoken shisetsu*) to the existing array of acute hospitals, geriatric hospitals, outpatient clinics, nursing homes, day care and home care services. The Geriatric Health Facility would take as inpatients transferees from geriatric hospitals (which have lower skilled staff ratios than acute hospitals), and would treat as outpatients impaired elderly in the community. From there, patients would either be returned to the community or sent to skilled nursing facilities. Geriatric Health Facilities would provide essential medical care, daily living support (meals, bathing, dressing, medications, etc.), and rehabilitative services (physical and occupational therapy). The complete array of services can be diagrammed as follows:

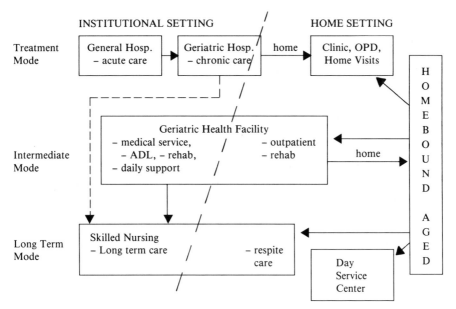

*(from Miyajima 1986: 37)*

In contrast to most Western countries, in Japan the family is by far the most common caregiver of the severely impaired elderly, probably accounting for three fourths of the nursing care to the disabled. A fourth of all primary caretaker-relatives are over sixty themselves (Maeda 1983), and the stress levels in many families are high. In one study (Nakajima, Saito and Tsukihashi 1982) 92% of caretakers of demented elderly said they felt some stress in their roles, 54% rated their stress as acute, and 50% said caretaking was a round-the-clock job. Respite care is the most frequently cited unmet need of the caretaker population. Yet without the free labor of these caretakers, Japanese society would be even harder pressed to meet the basic necessities of its aging population.

Given the unusual willingness of Japanese families to care for their elderly members, the progress of geriatric service development in Japan still bears close watching in the West. Continued urbanization, inflation, and increasing female employment, together with a falling birth rate, are eroding the ability of the Japanese family to care for its elderly at home. At the same time, the aged population is growing by leaps and bounds. Most developed societies are traveling along the same trajectory, albeit at a somewhat slower pace.

## A CRITICAL POSTSCRIPT

In three decades of interest, a Western image of Japanese aging has emerged. It is, of course, an image fashioned by the projects of gerontology, with its close ties to the social welfare and health professions. We have wanted to know how the Japanese cope with the "problems" of the aged – status, social integration, loneliness and boredom, financial and medical dependency. These are noble questions, and the answers to them may help us contribute to the overall health of our own aging societies. There are, I think, two dangers in this project as it is now being pursued.

First, there is the danger of false parallels. Japanese and American (and European) gerontologists are interested in many of the same questions, and learn much from one another's techniques and findings. The impressive flood of White Papers, Annual Reports, and survey research papers coming out of the Ministries of Health and Welfare and Finance, the schools of public health, and the Tokyo Metropolitan Institute of Gerontology (which publishes *Shakai Ronengaku* – Social Gerontology) tend to contain the same sorts of information one finds in Western gerontological research. But is it comparable? Does *netakiri* really mean "bedfast," or are both words really person-in-situation categories that include non-comparable aspects of physical and social environment? Japanese houses and Japanese attitudes toward illness and caretaking are on the whole hostile to both unstable ambulation and wheelchairs, but if this aversion could be overcome, would the definition of *netakiri* suddenly change? Other examples are abundant.

The second danger is that of culturalism. If things are apparently different in Japan, are they different because they occur in Japanese culture, or for some other reason? The cultural explanation implies resistance to change, whereas an economic or technical one may not. I often wonder whether 60% of the Japanese aged live with their children *largely* because housing is so expensive and age-appropriate environments are so scarce there. That explanation would make just as much sense as the cultural one; in fact I have heard an identical wish from the lips of older Japanese and older Americans –

the wish for housing that is "close but separate."

There is no simple safety from these dangers of interpretation. We can avoid them best by studying the relevant facts in their many explanatory contexts – cultural, biographical, historical and economic, macro-and micro-. There are as yet few descriptive studies of actual old people in their actual contexts, although Kinoshita (1984) and Campbell (1984) have made an excellent start. There are few life histories of ordinary old people, although Plath's middle-aged biographies are exemplary (Plath 1980). We need ethnographies of as many as possible of the huge variety of adaptations individuals and families have made to the "problems" of growing old in Japan.

This may seem an impractical suggestion, but a little imagination should reveal the extent of gains to be had from the attempt. We need ethnographies of interesting health care systems, like Sawauchi Mura and Musashino City (see Kiefer 1987). We need studies of institutions such as geriatric hospitals and nursing homes. We need descriptions of gerontological and geriatric professions. We need ethnographies of multi-generational families in various walks of life. We need age-focused studies of typical urban and rural communities, and those with high concentrations of aged. With these and other efforts, we can gradually build a fund of interpretive skill. It is only through direct observation that the webs of meaning emerge that show discrete facts as they are understood by those for whom they matter. It is only through the grasp of these relations that we can form robust causal models in which the cultural dimension takes its proper place.

## REFERENCES

Ariyoshi, S. (1972) Man in Ecstacy. Tokyo: Shinchosha.
Befu, H. (1962) Corporate Emphasis and Patterns of Descent in the Japanese Family. In Japanese Culture, Its Development and Characteristics. R. Smith and R. Beardsley, eds. New York: Viking Publications.
Benedict, R. (1946) The Chrysanthemum and the Sword. New York: Houghton Mifflin.
Campbell, J. (1984) Problems, Solutions, Non-solutions, and Free Medical Care for the Elderly in Japan. Pacific Affairs 57: 53-64.
Campbell, R. (1984) Nursing Homes and Long Term Care in Japan. Pacific Affairs 57: 78-89.
Clark, M. and B. Anderson (1967) Culture and Aging. Springfield: C.T. Thomas.
Cowgill, D. and L. Holmes (1972) Aging and Modernization. New York: Appleton-Century Crofts.
DeVos, G. and H. Wagatsuma (1973) Role Narcissism and the Etiology of Japanese Suicide. In Socialization for Achievement. G. DeVos and H. Wagatsuma, eds. Berkeley and Los Angeles: University of California Press.
Embree, J. (1939) Suye Mura: A Japanese Village. Chicago: University of Chicago Press.
Hashimoto, A. (1986) Formal and Informal Support Systems in Perspective: Japan and U.S.A. Paper presented at XI World Congress of the International Sociological Association, New Delhi.
Hearn, L. (1920) Japan, An Interpretation. Boston: C. T. Tuttle.
Ikegami, N. (1982) Institutionalized and the Non-Institutionalized Elderly. Social Science and Medicine 16: 2001-2008.
Imamura, A. (1987) Urban Japanese Housewives: At Home and In the Community. Honolulu: University of Hawaii Press.
Inouye, Y. (1982) Chronicle of My Mother. Tokyo: Kodansha.
Kiefer, C. (1974) Changing Cultures, Changing Lives. San Francisco: Jossey-Bass.
Kiefer, C. (1987) Care of the Aged in Japan. In Health Illness and Medical Care in Japan. E Norbeck and M. Lock, eds. Honolulu: University of Hawaii Press.
Kiefer, C. (1990) The Elderly in Modern Japan: Elite, Victims, or Plural Players? In The Cultural Context of Aging. J. Sokolovsky, ed. Wesport: Bergin and Garvey.
Kii, T. (1979) Recent Extension of Retirement Age in Japan. The Gerontologist 19: 481-486.
Kinoshita, Y. (1984) Social Integration in a Japanese Retirement Community. Unpublished doctoral dissertation, University of California, San Francisco.
Kleemeier, R., ed. (1961) Aging and Leisure. New York: Oxford University Press.
Koyono, W., H. Shibata, H. Haga, and Y. Suyama (1986) Co-residence with married children and health of the elderly. Shakai Ronengaku 24: 28-35 (in Japanese).
Lebra, T. (1984) Japanese Women: Constraint and Fulfillment. Honolulu: University of Hawaii Press. Maeda, D. (1975) Growth of Old People's Clubs in Japan. The Gerontologist 15: 254-256.
Maeda, D. (1982) Family Care in Japan. The Gerontologist 23: 578-583.
Maeda, N. (1983) Health Schemes for the Aged in Japan. Scientific Session Papers, Ninth Joint Tokyo/New York Medical Congress: Tokyo.
Ministry of Health and Welfare, Japan (1987) Health and Welfare Statistics in Japan, (1987). Tokyo: Health and Welfare Statistics Association.
Miyajima, S. (1986) Combatting Elderly Disability. Kosei no Shihyo 10: 31-39 (in Japanese).
Nakajima, K., K. Saito, and Y. Tsukihashi (1982) Actual Conditions of the Demented Elderly and their Families. Hokenfu Zasshi 38: 10-47 (in Japanese).
Nakane, C. (1972) Japanese Society. Berkeley and Los Angeles: University of California Press.
Naoi, M. (1987) Work Career and Earnings of the Young Old. Shakai Ronengaku 25: 6-18 (In Japanese).
Niwa, F. (1962) The Hateful Age. In Modern Japanese Stories: An Anthology. I. Morris, ed.

Rutland, Vermont: C. T. Tuttle.
Norbeck E. (1953) Age Grading in Japan. American Anthropologist 55: 373-384.
Okamura, K. (1987) The Employment of Fixed-year Retirees: The Unemployed Situation and its Main Regulating Factors. Shakai Ronengaku 26: 3-17. (In Japanese).
Palmore, E. (1975) The Honorable Elders. Durham, NC: Duke University Press.
Palmore, E. and D. Maeda (1985) The Honorable Elders Revisited. Durham, NC: Duke University Press.
Plath, D. (1972) Japan: The After Years. In Aging and Modernization. D. Cowgill and L. Holmes, eds. New York: Appleton-Century-Crofts.
Plath, D. (1980) Long Engagements: Maturity in Modern Japan. Stanford: Stanford University Press.
Prime Minister's Office, Bureau of Aging (1980) The Present State of the Elderly. Tokyo: Ministry of Finance Printing Office (In Japanese).
Prime Minister's Office, Bureau of Aging (1982) Lives and Opinions of Old People: Report of a Cross-National Survey. Tokyo: Ministry of Finance Printing Office (In Japanese).
Smith, R. (1961) Japan: The Later Years of Life and the Concept of Time. In Aging and Leisure. R. Kleemeier, ed. New York: Oxford University Press.
Sparks, D. (1975) The Still Rebirth: Retirement and Role Discontinuity. Journal of Asian and African Studies 10: 64-74.
Sussman, M., J. Romeis, and D. Maeda (1980) Age Bias in Japan: Implications for Normative Conflict. International Review of Modern Sociology 10: 243-254.
Tozawa, G. (1986a) Present and Future Problems of the Health Care System for the Elderly. Kosei no Shihyo 10: 4-12. (In Japanese)
Tozawa, G. (1986b) Trends in Medical Costs of the Aged. Kosei no Shihyo 10: 13-18. (In Japanese)

ANDREA SANKAR

# 7. GERONTOLOGICAL RESEARCH IN CHINA: THE ROLE FOR ANTHROPOLOGICAL INQUIRY*

INTRODUCTION

In a review of the policies of developing nations concerning the elderly Treas and Logue (1986) identify four categories which describe the way different countries appraise the role of the elderly. In these the aged: (1) are seen as a low priority in development: (2) are cast as the consumers of scarce resources and as such impediments to development; (3) are portrayed as possible resources in development, as a flexible source of reserve labor in marginal enterprises; and (4) are depicted as the victims of the process of modernization.

The People's Republic of China (hereafter China) is perhaps unique among lesser developed nations in its awareness that social policy decisions regarding the aged cannot be made in isolation from policy decisions concerning the rest of the population. People's decisions about the number of children they will have, the kind of investments they will make, and, in some cases, the kind of social relationships they enter into are linked to their perceptions of their security in old age (Cain 1985; Davis-Friedmann 1983; Sankar 1981). Studying aging and the aged in China presents an opportunity for examining the elderly not only as a distinct group but also as an integral part of the social order, an integration and interdependence which is made explicit by current political, social, and economic policy. Further, within the Chinese system of bureaucratic allocation of jobs, chronological age itself, or better cohort, is a significant factor in determining people's experience and opportunity in life. Thus, the study of cohort inevitably raises questions concerning fundamental aspects of social life in modern China. And, of course, one studies old age in China for the compelling reason that it has one fifth of the world's elderly population (Ikels 1988).

Despite significant and serious arguments which attest to the need to study aging and the aged in China, the amount of research which has so far been completed is slim indeed. Much is currently underway and will be reviewed here as in-progress, yet the subject calls for far more attention than has so far been accorded it both in the West and in China. This paucity is in large part due to the considerable political restrictions which, until recently, have impeded all research, but especially ethnological and ethnographic investigation. Restrictions on Western scholarship, although more flexible than in the past, are still a force in discouraging the paid expansion of this research area. They

* This chapter is dedicated to those who died in the Tienanmen Massacre, Beijing, June 3-4, 1989.

offer a partial explanation for the thinness of our current understanding of aging and the aged in China particularly in regards to the contribution of Western scholars. (For ethnographic studies conducted prior to the Cultural Revolution (1966) or in Taiwan or Hong Kong see for example, Baker 1968; Cohen 1976; Fei 1949; Freedman 1958, 1966; Fried 1953; B. Gallin 1966; Gamble 1954; Ho 1962; Hsu 1948; Ikels 1983; Lang 1946; Lin 1948; Sankar 1978; Tawny 1932; A. Wolf 1968; M. Wolf 1972; C. K. Yang 1959; M. Yang 1945).

A more pressing impediment in our appreciation of aging and the aged in China, and one which is likely to grow more weighty rather than shrink with time, is the considerable diversity wrought by ethnic, regional, and economic differences. One cannot talk about the Chinese elderly with any real precision except when demographically describing them. For example, there is the fact that until very recently the population planning policy of restricting couples to one child did not apply to non-Han ethnic groups. As economic policies which favor coastal regions over the interior are pursued, regional economic differences and their effects on the well being of the elderly are likely to intensify.

This review has three main objects: 1) it will identify the kind of research which has been done; 2) it will synthesize what is currently known about the elderly and age in China with particular attention to those areas directly affected by governmental policy; 3) it will review research currently in progress and suggest areas which might be addressed in future work. It will take as its main focus the effect of state policy on the experience of age. In part this is an obvious orientation given the extreme social change which the Chinese Communist Party has fostered since coming to power in 1949. It also reflects the simple fact that until recently Western researchers had more access to policy pronouncements than to the actual situation and thus focused their research questions on issues related to state policy.

Despite the limits on our understanding of old age imposed by both the nature of the object of study and by the political context in which the study must take place, this is a particular compelling subject; for, it allows us to critically examine the almost mythical Chinese gerontocracy standing as it does for a "golden age" when families "really" took care of the elderly. This myth informs the policy ideals in the West and its critical examination is long overdue.

## SOURCES AND METHODS

To date, there has been relatively little ethnographic work focused specifically on aging and the aged in China, although some projects of this nature are currently underway and will be discussed in a later section of this paper. The ethnographic data presented here come from three sources. The initial results from Charlotte Ikels' work (1988, forthcoming) represent direct ethnographic data on Chinese elderly. Other data presented here are culled from ethnographic accounts which were not directly focused on the elderly but on related subjects such as life history or the family (Pasternak 1986; Huang forthcoming; Dong 1987; Davis forthcoming). They also come from Taiwan (Gallin 1986; Hsieh 1985; Kleinman 1980; Sando 1986) an area closely related to China but whose historical, cultural, and economic experience is sufficiently different not to justify inclusion directly within the topic even given the great diversity included under aging and the aged in China. I have chosen not to review the material on Hong Kong because its predominantly urban and industrialized population, a significant proportion of which is made up of refugees from China, makes it difficult to compare it to the mainly rural population of China. There are, however, potentially strong similarities as Ikels (forthcoming) points out. Contemporary urban caregivers in China are in similar position to their Hong Kong counterparts fifteen years ago when the government used the tradition of filial piety to justify inadequate service provision to the elderly, leaving families to suffer and sometimes collapse under the strain of full-time parent care.

To provide the reader with the background necessary to fully appreciate the salient issues in Chinese gerontology, I have had to draw extensively on material which ranges from interviews (Davis-Friedmann 1983, 1985b; Davis 1986; Greenhalgh 1986; Ikels 1983, 1988, forthcoming; Whyte 1988), to a combination of interviews, participant observation, and questionnaires (Croll 1987; Davis forthcoming; Dong 1987; Huang forthcoming; Nee 1985, 1986; Pasternak 1986; Wolf 1986) to survey research in China (Arnold and Zhao, 1986; Gui *et al.* 1987; Liu and Yu 1988; Yu *et al.* forthcoming), and analytic pieces based on secondary sources (Davis-Friedmann 1985a; Keyfitz 1984; Liang, Tu, and Chen 1985; Liang and Gu n.p.; Olsen 1987, 1988).

Before reviewing this material, it is important to again emphasize its inherent limitations. China is a large country with distinct ethnic and regional divisions. Even the results of a random sample of 1000 respondents from an urban area must be viewed with caution because the results represent, at best, the conditions of that urban district, or perhaps of only one area of that district in a city such as Shanghai. In some respects the qualitative data provide a more accurate picture because the reader can clearly appreciate the nature of the study population and hence its limits. The portrait which follows will be painted with broad descriptive strokes. It should, if successful, raise more questions than it answers.

## THE EXPERIENCE OF AGE IN CHINA

### THE IMPACT OF STATE POLICY

Since the 1949 Revolution, The Chinese Communist Party has attempted to institute social changes which extend far beyond those tied to the means of production. The most significant policies in this regard are those pertaining to population control, or the lack of it, and to the distribution of jobs. The ramifications of these policies have had a profound impact on people's experience of age.

*Demography*

According to the 1982 census, there were 49 million people 65 and older in China. Although the age of 65 is used in the United States for the purposes of long range planning, in China the general retirement age is 60 for men and 50 or 55 for women depending on the type of job they held, i.e. the number of those 60 and over will be cited as the relevant statistic because it is the one on which the Chinese base their planning. In the 1982 census there were 76.6 million people sixty and over who constituted 7.6% of the population (Liang, Tu and Chen 1985). According to this census, the life expectancy for women born in 1981 was 69.4 years and for men 66.4 years. Given the decline in birth rate and the steady mortality rate the population is expected to age rapidly especially if the one-child fertility policy, which is designed to reduce the population size is successful. Projections concerning the future growth of the elderly population vary somewhat according to different forecasters. The above authors suggest that there will be a relatively favorable dependency ratio until the late 2020's when the large birth cohorts of the period 1960-1975 begin to retire. At this point the proportion of elderly in the population will begin to increase rapidly, reaching 20% by 2040. Other researchers estimate that a crisis will be reached as early as 2000 (Bannister 1987) or 2005 (Keyfitz 1984).

These figures are slightly misleading because the population will age at different rates depending on its geographic location and urban/rural mix. In Shanghai those 60 and older already account for 11% of the population. This figure rises to 13% if the suburban regions are included (China News Analysis 1984). There, it is projected that by the year 2000 the figures will be 18% for the urban area and 20% for the urban and suburban areas combined. Combining urban and suburban areas does not always result in an increased ratio. In Tianjin those 60 and older currently account for 8.7% of the population, a figure which drops to 7.97% when combined with the suburban population. This brief comparison serves as a reminder of the dangers of making global statements when one is describing the aging population in China.

Demographers speculate that some kind of crisis will develop in the first half of the twenty-first century, brought on by the movement of the huge birth cohort from the 1960-75 period into retirement, and the introduction of a strict population planning program which aims to develop negative population growth through the one child family campaign. If this campaign is completely successful, Bannister (1987) predicts that the elderly will constitute 41-45% of the population by the middle of the next century. Liang, Tu and Chen (1985) estimate that by 2042 the dependency ratio will be 34 elderly for every 100 workers. The government and the research institutes which sponsor research by Westerners will expect that much of the research on aging carried out in China during the coming years will address this phenomenon in some fashion (Greenhalgh 1988).

*Age Stratification*

Government policy has had a powerful and pervasive impact on the experience of age through bureaucratic allocation of jobs which creates strikingly different opportunities and rewards for individuals depending on the cohort into which they were born. In this respect, argues Davis-Friedmann (1985a), China is similar to other contemporary state socialist systems. She describes how a system of age stratification has developed whereby "first-comers", those entering the workforce during a period of economic expansion, are systematically given advantages over "late-comers" those who arrive on the job market during a time of economic contraction. Because of consistently high birth rates and uneven rates of job creation, the time at which one entered the work force determines one's access to a vastly different set of opportunities and rewards in life. Thus, a system of generational cleavages has evolved whereby many newly retired people have a higher income than their middle aged children.

Davis (1988) attributes the development of this system of job allocation to the strength of an entrenched bureaucracy, to the control of the military over the bureaucracy for many years (i.e. in the military the needs of the labor force for personal fulfillment and advancement need not be attended to), and to the manipulations of the political leadership in its move towards transforming the economy. Since the revolution class inequalities have decreased only to be replaced by inequalities based on age, gender, region, and economic sector. These inequalities are likely to increase as the government moves towards contract labor as the officially sanctioned mode for recruiting new labor. This allows managers to have more flexibility in managing their workforce by doing away with the "iron rice bowl", the job for life system. Contract workers, however, have a distinctly reduced system of rewards and opportunities than permanent workers, those hired before 1986 when the system of contract labor took effect. The potential antagonisms between the generations which this could create worries many political planners and was noted repeatedly in the works of Chinese scholars in the Gerontological Society of America's special publication, *Aging China: Family, Economic, and Government Policies in Transition* (1987).

*The One-Child Campaign*

Perhaps the relationship between the welfare of the elderly and the family planning activity of the young has never before been so explicitly recognized as by those government planners trying to make the one-child campaign a success. A common bleak image used to criticize the campaign's goals is that of a middle aged couple responsible for the care of four elderly parents and one teen-ager. To disarm this fear and alter the socioeconomic factors which lead to high fertility, family planning programs have begun to emphasize the need of the government to "render service" (Greenhalgh 1986). Cadres, the professional political workers and members of the Communist party, have become involved in helping couples solve problems ranging from old age support to gaining access to labor and capital. Davis (1986) argues that the specter of the overwhelmed middle aged couple is inaccurate. She points out that in the near future, with the movement of the baby-boom generation through the workforce, dependency ratios will drop. This will produce more spendable income. A whole range of social shifts may result. More income could reduce the need to share housing, although the number of available units remains a problem. It could also allow the middle aged to hire help to care for the elderly. She speculates that the reduction in the necessity for intergenerational interdependence will produce new forms of family relationships, similar to those that exist in Western industrialized countries, which demonstrate the full spectrum from great interdependence to complete dependence (1986).

Aware of the problems the one-child policy will create for the family in its ability to care for the elderly, the Chinese government has already begun to modify the campaign. In rural areas numerous exemptions, including those designed to acknowledge patriarchal traditions, have been introduced. In effect, these allow for most families to have two children. In urban areas, however, the fertility rate is between 1.0-1.4 births per woman which will cause severe population aging unless fertility increases or there is significant in-migration (Bannister 1987). Several solutions involving birth spacing are being considered to ameliorate the problem of urban aging. One possible solution would be to adopt a policy which allows a couple to have one child soon after marriage and a second four to six years later (Greenhalgh and Borgaarts 1987).

The one-child campaign poses a significant challenge to the government to provide for the pension and medical coverage for workers after the low fertility rate has weakened the family structure on which they now depend (Bannister 1987).

## SOCIAL CHANGE AND THE ELDERLY

Beyond specific governmental policies, the state has affected the experience of age through the wider social transformation which it has fostered. The impact here is seen in such areas as the indirect influence of housing availability on family residence patterns, efforts to discourage son preference, and the kinds of reciprocal relationships between generations which urban living and the employment situation necessitate.

*Co-residence*

Chinese elderly continue to reside with at least one child in significant numbers. Davis-Friedmann (1983) in a study undertaken in the 1970's estimated that at that time 70% of rural elderly and 90% of urban elderly lived with at least one child. These figures have changed slightly according to a more recent study which draws on the 1982 census. Gui *et al.* (1987) report that in Shanghai three out of four elderly live with at least one child. In the most recent available results, Liu and Yu (1988) report, based on a random sample of 5050 people 55 and older living in Shanghai, that 73% of the sample live with at least one child and only 6% live alone. Although co-residence reflects the cultural ideal of the extended family living under one roof, more practical factors may promote the prevalence of this pattern. The high rate of co-residence, especially in urban areas, may in part be due to a severe housing shortage. Davis (1986) suggests this when she compares the higher co-residence rate in Shanghai, with its extreme housing shortage, to the lower rates of both Beijing and Tianjin, which have recently increased their housing stocks.

In urban areas joint living also may be related to household wealth (Davis 1986). Davis's 1986 research in Shanghai indicated a curvilinear relationship between household wealth and family size. The wealthier and the poorer families included three generations while those in the middle had one or two. Only the wealthiest, usually veteran cadres, have generous living quarters and the poorest simply jerrybuild extra spaces onto their living quarters to accommodate married children.

In her ethnographic and interview study of the families of researchers at the Chinese Academy of Sciences in Shanghai, Dong (1987) substantiates Davis's analysis, at least for the wealthiest group. The difference in living space allocated to the older, established scholars and the middle aged and younger researchers was dramatic. As researchers progressed in seniority and authority within the Academy, they were given more living space in accord with their rank. Thus, senior researchers with grown children not residing at home might have a four room apartment with up to 92 square meters of living space. On occasion this space had to be shared with other families which the work unit would temporarily house there. But this spacious allocation was the norm for the senior researchers and their spouses. In 1980 the average Beijing resident had 4.79 square meters (52 square feet) of floor space in which to live. Among the junior researchers at the Academy, it was not unusual for a couple and their one child to live in 12 square meters. When a couple did not have sufficient living space, one or more children were sent to grandparents to live until the family obtained adequate housing. With such significant disparities in living conditions it is not surprising that many young couples ended up living with one set of parents.

Research results are not consistent in reporting the preferences of the elderly concerning co-residence. According to Yuan (1987) the shift toward more independent residence in Beijing corresponds with the desires of the elderly. In reporting results of a survey from Beijing, he states that 100% of the elderly interviewed who were living alone wanted to continue to do so and 23.5% of those who were currently living jointly would prefer to live only with a spouse. In the rural areas the situation seems somewhat reversed. Results from a survey of rural areas of Heilongjiang and Sichuan showed that 40% of the elderly reported that living jointly was the ideal but that only 27.9% actually did so (Yuan 1987). In a survey of elderly Shanghai residents, Gui *et al.* (1987) found that 32.7% preferred to live only with a spouse and with a child nearby. Significantly, they report that the percentage of those preferring to live alone declined with age. A 1986 study conducted in Hubei, which does not indicate its urban/rural mix, found that 48% of those elderly living with their children cited dependency as the reason, 23% identified helping with the chores, 15% responded enjoyment of family life, and only 14% gave as the reason housing shortage (Liang and Gui n.p.). Because of the dependence of many elderly, especially those seventy-five and above, on their family for support and because of the severe housing shortage in most urban areas, it is difficult to clearly establish their preference for living arrangements when alternatives to their present situation are non-existent.

Pasternak (1986) reaffirms the difficulty of imputing values or attitudes about co-residence from housing patterns. In a study on demographic trends conducted in two neighborhoods in Tianjin in 1981-82, Pasternak reports that neolocal residence has been increasing since before 1949. Migrants coming to the city accounted for the early trend. Except for times of economic or political reversal, the trend has continued. Since 1969 shortages in housing stock have put a minor restraint on neolocality. Not only are housing units in short supply but the space within a unit is extremely limited forcing young married couples to leave their parent's quarters as soon as possible. According to Pasternak's informants this is not always desirable.
They cite a preference for living with parents in part because of the help provided by the mother-in-law in child care and household management. Older people, however, appear to be less enthusiastic about residing with their children. Pasternak reports that a 1985 survey of 1000 households showed a 10.1% decline between 1983-85 in those desiring to live with their children to only 50% ("China Daily News", 2 January 1986:4, cited in Pasternak 1986:34). Despite this stated preference only 4% of his informants lived alone; 49% lived in nuclear households and 45% were members of extended households. Clearly more work needs to be done to disentangle the effects of the housing shortage from residence preference.

*Patrilineality*

Sons and daughters are equally required by law to support their elderly parents. But the cultural expectation that one son will assume the care of aging parents remains strong. When parents do reside with their grown children, there is a clear preference for residing with a married son. In the study of Shanghai, Gui et al. (1987) found that among those elderly who preferred living with a child as opposed to living alone with a spouse, 26.9% preferred living with a married son, 18.1% had no preference and 12.4% preferred living with a married daughter. Davis (1986) found fewer than 10% of her older informants living with newly married daughters; in contrast, 40% of the newly married sons continued to reside at home.

The continued strength of son preference has become strikingly clear with the introduction of the one-child campaign. Despite the government propaganda which features a girl child as the one child, there remains a strong preference for sons, so much so that the practice of female infanticide again appeared after the initiation of the campaign (Nee 1986). As the political pressure for gender equality waned with the move toward decollectivization, the authorities were unable to discourage the expression of the preference for sons, especially in its extreme form, female infanticide. The one-child campaign in rural areas has been modified to acknowledge the patrilineal tradition by allowing families to have a second child if the first is a girl (Greenhalgh 1986). In this way, if both children are girls, one daughter can be married out and an uxorilocal husband can be married in to care for the parents in old age.

Davis-Friedmann (1985b) suggests several economic and structural reasons for the continued preference for sons in rural areas. Unlike urban areas where children are often seen as long term dependents, in rural areas elderly parents rely on their children for support in old age. She attributes this to three factors: (1) because women do not earn as much as men, a disadvantage which is compounded by the unpaid housework which they must perform, they cannot provide as well for their parents, (2) although the government has tried to discourage surname exogamy, the continued strength of this practice makes it difficult to maintain close ties with married daughters, (3) the practice of ancestor worship in which the son must officiate in the rites for the deceased parents creates a religious preference for sons.

In an impressive example of ethnographic work, Davis (forthcoming) explores the significance of domestic interiors and of popular culture in Shanghai. Based on a sample of one hundred families containing a woman born between 1925 and 1935, she argues that the concept of the "mother's house" serves as a strong emotional anchor for urban informants. She analyzes photographs in the homes she visited and argues that matrilineal ties are very strong despite the pattern of patrilocal residence. "Home is the home of the mother," she concludes. The interiors of her informants' homes, despite their patrilocality, clearly reflect the strength of the tie to the mother and the primacy of women in the domestic sphere. To some extent this is what one would expect in a strongly patrilineal society. This finding, if born out in large data sets, may indicate the kinds of caregiving relationships to which older women might have access.

## Reciprocity

Relationships within the Chinese family have traditionally been marked by strong intergenerational support and reciprocity. Children care for their parents in old age in return for their care when young and for the gift of life. Few families have the surplus wealth to allow the elderly to "retire" within the family. Even in extreme old age the elderly are expected to contribute to the household economy whenever possible.

The way such support is actualized differs markedly in rural and in urban areas. In rural areas where 75% of the working population lives, the elderly have been cared for by their sons, who consider it a joint responsibility. At the time of household division, either one son is given a larger share of the patrimony with which to provide the aging parents with care, or the sons divide up the support for the parents more or less equally. In some cases different children care for the father and for the mother.

In the 1950's, 60's, 70's, and even into the early 80's agricultural production was collectivized. Neighbors worked in work teams; teams were formed into brigades (which were usually synonymous with the former village); and brigades were organized into communes. Workers were assigned work points according to age, gender, and the amount they worked, not what they produced. At the end of each year the work points were totaled and people were paid in food stuffs and cash according the number of points they had accumulated. The elderly who had retired from the workforce were still able to contribute to the family income. They were not obligated to work a set number of days on the collective and could instead concentrate their efforts in the private sector and contribute to their own subsistence and sometimes to the household economy. Davis-Friedmann (1983) illustrates in detail how the raising of a pig to be sold on the private market by the grandmother could generate substantial income for a family.

The potential economic contribution of the elderly has increased under certain circumstances since decollectivization and the revival of traditional family production units. In rural areas it was often the "young old" men in their fifties who initially benefited from the introduction of the new responsibility system (Davis-Friedmann 1985a). They had the developed social networks, skills in traditional handicrafts, marketing savvy, and often the ownership of traditional means of production such as specialized artisan tools. Even a much older person who is skilled in handicrafts or can free up the labor of adult family members by attending to household chores can aid the family in advancing economically in the now tightly competitive atmosphere of rural production where even marginally useful labor can give a family a competitive advantage.

The rural elderly without sons to care for them, those eligible for the "5 guarantee" program have not fared as well under the new responsibility system. Without the social and political pressure to contribute to the social welfare fund, more and more elderly without sons must care for themselves and face a significant decrease in standard of living. In Huang's (forthcoming) account of village life, the chief protagonist, P.S. Yie, first secretary of his brigade (a position roughly equivalent to a mayor), expresses serious concern for the effect of decollectivization on the poor and elderly. A decrease in the standard of living for the dependent elderly, those in five guarantee households has also been reported by Greenhalgh (1986). The new responsibility system may also be affecting the well being of elderly within their own families. In Davis's (1986) analysis, she reports that the new responsibility system has reduced the pressure on children to support elderly parents who can no longer make an economic contribution to the household. For example, Western researchers' reports show that between 10-20% of the elderly in some Guangdong villages are not living jointly with their children but must instead "go it alone."

The situation for the urban elderly is strikingly different. Due to a wage system which until recently has been strictly tied to seniority, senior workers by definition earned more than younger workers. For example, in Dong's study (1987) an average working couple at the Chinese Academy of Sciences in their 40's earned a combined income of 156 Yuan (approximately $78 US) a month, a couple in their 50's earned 231 Yuan a month, and a couple in their 60's 332 Yuan a month. (These figures do not reflect in-kind subsidies provided by the work unit). In 1978 the government decided to change the pension system in the state sector to increase access to jobs for younger workers. Since 1978 the government has pushed for the mandatory retirement of women over 50 in manual labor, over the age of 55 in white collar labor and for men over 60 which has resulted in a phenomenal growth in the number of pensioners (Davis 1988). Currently, pensions provide for 70% of the last wage. This generous retirement benefit has made these "young old" a significant economic force in their families.

The economic advantage for the newly retired has, in some cases, been compounded by those who choose to go into part-time or entrepreneurial activities. Participation in entrepreneurial activities appears to depend on the adequacy of retirement income. Liu and Yu (1988) report from their Shanghai study that less than 10% of those retired work full time and only 1.2% work part time. Of those retired, 88% did not want to work. Liu and Yu (1988) point out, however, the significant income differential between those fifty-five and older who are retired and those who are still working. The mean income for those working was 69.2 RMB and the median 67.00 RMB compared to a mean income of 99.8 RMB and a median of 93.00 for the retirees. The relative affluence of some elderly has not affected those elderly who did not work in state enterprises and therefore were not covered by a pension. They must still rely on their families and paid labor for most of their support.

As in rural areas, urban children are responsible for the care of their elderly parents. In most cases urban adult children continue to make regular contributions to their parents income. With a combination of a high retirement income and possibly an ample apartment allocation, the urban young-old can be a potent force in their adult children's lives (Davis-Friedmann 1983) in some cases supplying support in the form of regular payments, free meals, or loans (Dong 1987). Liu and Yu (1988) report that elderly informants provided material support to their adult children from their pensions and that they received help in health care provision from the children. This also, as Dong points out, can lead to a reassertion of tyrannical patriarchal control over the lives of adult children.

The reciprocal relationship between the urban elderly and their adult children takes many forms. One of the most important roles which the elderly play is the provision of child care for working parents. Liu and Yu (1988) report that 50% of the respondents had grandchildren living with them and that 68% said they helped in child care. The extent of this care is not specified. Usually grandmothers play this role but grandfathers are on occasion the providers. Most adult women in urban areas work full time. The state childcare facilities will not take children under the age of two, and are said to be no where near adequate to meet the demand (Dong 1987). Some few work units provide on-site care for infants but these are rare and usually full. Grandparents, when available, are very important in the care of infants and young children. The grandparents may care for the child in the parents' living quarters, or the child may be sent to live with the grandparents, either the father's or mother's parents, until he or she can enter a day care center at age two or until school begins at age seven. In Dong's study the parents usually pay the grandparents for this care. The amounts reported by Dong were equal to or greater than non-kin child care. This is not always a satisfactory arrangement for either party. Parents would rather have the children with them; they worry that the grandparents are spoiling the children. The grandparents find the task sometimes overwhelming. An additional problem is related to task appropriateness. Dong reports that the newly retired, educated grandmothers are sometimes not willing to adopt the role of baby sitter. In one informant's family, the grandmother contributed to the salary of a governess to free herself from what she saw as inappropriate work. In this changing environment the new young-old can not be seen as an assured source of child care.

Elderly parents can also provide valuable assistance in household chores. With both members of a couple working full time six days a week, housework is extremely difficult to accomplish. This is compounded by a scarcity of modern appliances and in some cases even running water is lacking (Dong 1987). Household work also includes standing in long lines to shop, a daily task in the many households that lacked refrigeration. The nature of these difficulties is changing as more labor-saving appliances become available. Labor-saving appliances are not widely distributed, however, and the presence of a grandmother in the household is still considered to give a family a clear economic advantage. One of the most significant kinds of transfer from the older generation to the younger one was the practice of *ding-ti*. After the Cultural Revolution, young adults who had gone to the countryside to work with the peasants and to spread revolution, the "sent down youth", returned to the cities seeking work. To accommodate these youths and to ensure that they did not create social unrest, the government modified an established but highly restricted system of job transfer called ding-ti, whereby a parent, upon retirement, could pass his or her place in the work unit to a designated child. The modifications allowed any parent to exercise this option. In its first year of operation (1978-9) the retirement rate doubled and the number of retirees increased 400 per cent between 1978 and 1985 (Davis 1988). This benefit is being phased out because of the management problems it created although it continues to exist informally (Dong 1987).

## HEALTH, ILLNESS AND CARE

There is little systematic epidimeological data describing the health and functional status of the Chinese elderly. Studies to assess this population are underway as are studies to determine how long term care is provided both within the family and from the outside. The data presented here represent, with the exception of the Alzheimer's study, snapshots of the situation. Considerably more data is needed.

### Health

The government has established some policies which are aimed at enhancing the quality of life of the elderly thus indirectly reducing their burden on the family by improving health and well being. National magazines written specifically for the elderly and several minor publications and political pronouncements aimed at the elderly urge them to remain physically and socially active. Numerous educational programs are available to retired workers (Hu, Wang and Zhang 1987). In some urban areas, residents committees have set up a kind of senior center *(lao ren zhi jia)* where they read, meet, and play games and urban elderly are encouraged to volunteer their services on behalf of political and social causes (Olson 1988).

### Illness

Yu *et al.* (forthcoming) have conducted the first longitudinal study of Alzheimer's Disease and dementia in China. They tested a probability sample of 5,055 non-institutionalized elderly 55 and older in Shanghai with the Chinese version of the Mini-Mental Status Examination (CMMSE) which they developed. In their initial findings the most remarkable results are the relative well being of the sample and the gender differences in illiteracy and cognitive impairment. Overall 4.1% of those 55 and older may be classified as having severe cognitive impairment while they found 14.4% with mild impairment. The rates for women were consistently higher than for males by a ratio of 3.75 in the severe group and 2.6 in the mild group within each age division. The cognitive impairment rates vary by education within each age group. When sex is controlled for, educational attainment had a highly significant inverse relationship with the prevalence of cognitive impairment. When education was controlled for, sex was significantly associated with the prevalence of cognitive impairment. Their findings suggest that basic educational deficits can have a significant impact on human cognitive functioning.

There is the important but relatively untouched area of the elderly's perception of health and illness. In a tangential way, Kleinman's study of culture, health and illness in Taiwan (1980), pointed out that the elderly in many cases have a complex, traditional understanding of the relationship between old age and disease. Sankar (1984) described how this conceptual relationship affects both the decision to seek medical care and the care which is expected and accepted.

In a study of two neighborhoods in Guangzhou, Ikels (1988) includes in the reporting of her results observations which are relevant to this topic although not of central concern to her research. Ikels (1988) suggests that disability rates reported by her elderly informants are lower than they might otherwise be given the extent of physical impairment because of co-residence and family organization. That is, by continuing to live within the family the elderly may not need as high a level of functional ability to remain active as someone who resides alone which is the preferred residence pattern in the West. This is an extremely interesting suggestion and one which needs more systematic investigation.

Without attempting to adequately discuss the field I will mention four of the more prominent works by Chinese authors in the area of health and well being. In a review of the chronic health problems of the old *Laonin Baojian Zhishi* (Information for Preserving the Health of the Old; 1978), Xia Lianbo describes the different problems which can be helped by diet and exercise. This popular book is one of the first to define old age as a social problem.

A volume edited by Yuan Jihui called *Laonian Wenti* (Problems of Old Age 1986) includes essays based on recent fieldwork on such topics as elder suicide, remarriage, economic status, care of the childless, recreation centers, psychiatric needs. Its appendix lists key recent publications on issues concerning the elderly. Yuan Jihui has also edited the two volume *Laonian Shenghuo Yanjiou* (Research on Urban Elderly 1986). The first volume reviews the 1982 census and the 1983 survey of elderly in Shanghai with special attention to health status, family relations, and economic resources. The second one further discusses the 1983 survey and provides more information which compares rural and urban elderly in the Shanghai area.

*Single and Childless Elderly*

Studying the lot of single and childless elderly provides a revealing perspective on the formal and informal provision of social welfare in a society. In pre-revolutionary China there were a myriad of cultural forms to allow for the provision of care to elderly without children. In a study of the remaining members of an anti-marriage movement in Guangdong, Sankar (1978, 1981) documented numerous forms of social relations which enabled childless elderly to be cared for in old age. These included from the creation of fictive kin relationships which established ties of mutual obligation, formation of communal associations among childless elderly, the adoption of a relative's child, care by the lineage, entrance into the religious life, and contracts which arranged for one man's wife to be "rented" for a specific time period by another man for the purposes of bearing children who would then belong to the lessor and his wife. Davis-Friedmann (1983) recounts the strategy of childless rural elderly, male and female, in post-revolutionary China who create strong reciprocal bonds of mutual obligation with younger neighbors, usually "sent down" urban youth, in return for care in old age. These bonds were necessary despite the assistance provided by the Five Guarantee program which was administered unevenly and provided only minimal help in housing, clothing, medical care, education and funeral expenses.

Since decollectivization these childless elderly are some of those at greatest risk. In 1986 there were 347,979 needy childless elderly in urban areas and in rural areas there were 2.4 million Five Guarantee households of childless or single elderly (Liang and Gu n.p.). In explaining this reduction in care Nee (1985) speculates that one of the reasons that collectivization did not work was because of the fear of "free-riders", those who did not contribute their fair share to the collective effort but who were entitled to an equal share of the profit. He suggests that the negative experience with "free-riders" under collectivization has eroded a sense of community charity leaving those who used to be assisted by the Five Guarantees, which included the childless elderly, more vulnerable to economic reversals than they were in the past.

This problem is by no means being ignored by the government. If childless elderly meet the qualifications of the "3 No's" – no children, no capability to work, and no other means of support – they will be eligible for social welfare (Liang and Gu n.p.). The government has chosen to make the childless elderly the target of social welfare programs for two reasons, Ikels argues (1988): first, they are a small and limited number and therefore the cost of caring for them cannot get out of hand; second, adequately providing for needy elderly will help promote the one-child campaign. Government planners eager to make the one-child campaign a success realize that the specter of elderly, no matter what their family composition, who are clearly destitute will confirm people in their desire to have more children. In some areas the local cadres have increased the amount of assistance to the Five Guarantee elderly, assigned people to help them, and constructed old age homes furnished with modern conveniences. In other areas, worker-contributed pensions have been instituted for childless rural elderly (Greenhalgh 1986).

*Long Term Care*

Liang and Gu (n.p.) estimate that at the present time there are some 12 million disabled elderly requiring assistance in daily life. Only .05% of them are cared for in an institutional setting. It is difficult to conceive how this ability to care for the elderly outside of institutional settings will continue with the decrease in family size. Even if the one-child policy is modified to allow two children and the proportion of elderly only reaches 22% by the middle of next century, the numbers of the oldest old, those 80 and above will still continue to increase rapidly and will create an immense social and economic burden (Bannister 1987). It is further likely that the problems created by shifting demographics will be exacerbated as the new economic freedom increases employment choices for rural youths, attracting them away from the farm and away from the care of aging parents (Ikels n.p.). Families are developing a variety of informal solutions to these problems, while the government is moving to address the issue both directly and indirectly. If the government were to relax its policies discouraging in-migration to the cities, this might alleviate the urban family's care burden by making hired help available (Bannister 1987).

The concept of care for the elderly outside the boundaries and idioms of kinship is slowly emerging. For families that are separated and for others where both people in the couple work full time, private home care is becoming an option. In recent years there has been a rapid growth in the number of young women from rural areas who migrate to cities to work as maids (China News Analysis 1987). These maids, sometimes called *baomu* (nursemaids), in many cases appear to serve the role of home health aide (Dong 1987). They come to achieve independence from their families or to save money for a wedding. In some areas there are government initiated home care services (Fang 1987). Their pay varies according to skill but can go as high as 200 yuan a month, which is comparable to the lower level researchers at the Chinese Academy of Sciences. However, it is difficult to find and retain *baomu* (Ikels 1988).

The government has developed a number of policy initiatives designed to indirectly support the family and alleviate the strain of caring for the elderly. The policy of assigning college graduates who are only children to a job near their parents is an example of this. Because university graduates must accept any job assignment given them at graduation, and because these job assignments need not be near their home town, young, highly qualified students are choosing a local college rather than a national university for their advanced training so as to ensure that they will remain near their parents. As a result, some university candidates seek education which is not commensurate with their skills according to Dong (1987). To avoid this and to promote the one-child family, the state is trying to settle only children near their parents and is giving the aim of reuniting a family high priority in approving job transfers (Dong 1987). Yet the problem still remains.

In addition to the informal solutions to parent care, the government is moving towards the actual provision of care by building old age homes. These are still few in number. In 1984 there were 169,000 or .33% of those over 65 in nursing homes (Liang, Tu, and Chen 1985). This represents 684 homes in urban areas caring for 22,000 people and 11,000 homes in rural areas caring for 147,000 people. The quality of these homes appears to some extent to be tied to the local family planning policy. In areas where this is a high priority the homes are appealing, representing a visible symbol of the state's attempts to reassure people about their care in old age (Greenhalgh 1986). They are not nursing homes, however, but homes for the childless elderly capable of self care. Eventually they will take on more of the aspects of a nursing home as their inhabitants age. The government has set a goal of establishing at least one old people's home in each township (*xiang*) (Olsen 1988).

The government is encouraging the development of pensions in the countryside. But the progress is slow. In 1980 only 180,000 or approximately .4% of the rural elderly received pensions. By 1983 that number had grown to 600,000 with approximately 9,000 brigades providing some kind of pension (Liang, Tu and Chen 1985). This issue of rural pensions is a significant one for China's future development. It is one, however, which will be extremely difficult to resolve given the potentially large numbers involved and the limited resources with which planners must work.

Direct services are available for childless urban elderly from the social welfare fund. The government also subsidizes the home services provided by home care teams to the most debilitated elderly who require complete care (Liang and Gu n.p.). Neighborhood clinics operate a Home Sickbed program for people unable to get to the clinic. Physicians attend to those recently discharged from the hospital, acute and emergency cases; and nurses visit the chronically ill (Ikels 1988).

*Death*

Although death is not the special domain of the elderly, it is a topic which greatly concerns them and their families. Traditionally providing for an appropriate and proper funeral was one of the most important obligations which a child owed its parents. As parents aged both they and their children were aware of the need to provide an adequate funeral. In many cases the parent saved money to ensure the adequacy of the funeral. With its attempt to change mortuary ritual, the government has directly intervened in an area of central concern to Chinese elderly.

## Mortuary Ritual

In a piece based on interviews with refugees and on secondary sources, Whyte (1988) explores the recent history and current status of death in China. He reviews the various government campaigns to limit expressions of mourning and to introduce cremation as the preferred alternative for disposing of the corpse. These proposals appear to have taken hold in the urban areas. In the rural areas, after a period of observing the ban on ostentatious displays and having burial sites situated on arable land, these practices are beginning to return.

The changes in funerary rites which have occurred in urban areas appear to be significant. There the deceased is usually mourned in public rites conducted by the work unit which addresses the deceased's contribution to society. This contrasts to the traditional private rites which stressed the afterlife. As a consequence of the changed focus on positive contribution of the deceased, death is not seen as polluting nor are those associated with the death polluted, except in the case of a bad death like a suicide. Another significant change is the lack of sharp distinction made between mourners who are patrilineal kin and others or between sons and daughters.

Whyte also concludes that although the transformed urban funeral rites convey a different meaning, they continue to represent much of the traditional ideology, e.g. the continuity between this world and the next, the responsibility of kin to find a proper resting place for the deceased's remains, and the influence on the living which the failure to attend to these matters can have.

## THE FUTURE OF GERONTOLOGICAL RESEARCH IN CHINA

### Work in Progress

The current research which is now being carried out in China, whose investigators have consented to provide descriptions for review, represents in-depth, theory-generated and analytical approaches to significant research problems. Roughly four kinds of questions are being addressed. The first is the impact on the elderly of the process of economic change, that is a re-examination of the "modernization" argument in light of the Chinese case. The second is an investigation of the role of a harsh environment on physical aging. The third is an examination of life cycle development in the context of significant social change. Finally, there is work that analyzes the dependency relationships involved in caregiving. The first three, especially the first one, explicitly focus on change, the fourth implicitly acknowledges it with the rapid growth in the elderly population used as a rationale for the study. With the exception of the study by Dong (1987), who attempted to develop a theoretical framework analyzing the relationship between state policies and family life and

Davis (forthcoming) who utilizes theories of meaning and context in her analysis of popular culture, these studies are rather more theoretical than those previously conducted. They promise to link the field of Chinese gerontology to major analytic and theoretical trends and to establish a solid basis for cross-cultural studies.

An intensive ethnographic study of two villages in Zhejiang Province is being carried out under the direction of Melvin Goldstein and Charlotte Ikels. The purpose of the project is to examine aging in a non-Western society undergoing the first stages of massive program of economic modernization. The theory guiding this undertaking is that the decrease in the status of the elderly which is commonly associated with the process of "modernization" is instead the outcome of specific conditions such as the elderly's loss of control over productive resources and/or the general level of poverty in a society which may force middle aged adults to allocate scarce resources to their children rather than their parents. Further, scarce economic conditions could encourage younger people to migrate out of an area and leave their elderly parents behind (Ikels, personal communication). They will examine the role of these specific factors as well as assess the role of traditional values such as filial piety in the fate of elderly rural Chinese under the responsibility system.

The research design will involve intensive participant-observation in two villages, one which has prospered under the new system and one which has not. As such it will be the first sustained ethnographic examination of the elderly in China. The design utilizes a kind of quasi-experimental approach in which participant-observation, interviews, historical record review, daily household activities and exchanges as well as survey data are combined to create two case studies which will then be compared to detect differences and to determine the factors which account for the differences in the status of the elderly in the different villages.

The design should allow the researchers to determine the manner in which economic and social changes affect access to and control over resources. Precise, in-depth data of this sort should help move the debate about the effects of modernization away from the examination of end results and toward a study of the process and forces which create those results. More specifically, village ethnographies conducted during this period will be extremely beneficial in documenting the extent to which the social welfare system which operated during the collectivization period has been dismantled and the effects this has had on the health and welfare of the elderly. Participant-observation will be particularly important not only for the processes and effects it will identify and document but also because it can elucidate the ways in which people's understanding of the social world and the events surrounding them affect family and community dynamics.

Melvin Goldstein and Cynthia Beall, in collaboration with the Tibet Academy of Social Sciences in Lhasa, are conducting a project that examines how the dissolution of the communes has affected the elderly nomads in the Tibetan Autonomous Region. It will compare the status of the elderly during three periods, the traditional society (pre-1959), the early revolutionary period and the Cultural Revolution (1959-1981) and the post-Mao era of the new responsibility system (1981-1988). This study will also examine the physiological adaptation of the elderly to one of the harshest physical environments in the world. The completion of this study and the one by Goldstein and Ikels should allow this group of researchers to compare and contrast the Tibetan situation with that among the Han populations and to make a significant theoretical contribution to the debate concerning the status of the elderly in the modern world (Goldstein, personal communication).

Ikels is also involved in an intensive interview study designed to determine the impact of the needs of a rapidly aging population on society and on the family. (Preliminary results from this study were presented in the previous section.) She assessed the functional ability of the elderly using an assessment protocol of her own development, determined the expectations of the elderly and their family members as to patterns of care for the disabled elderly, and documented the types of care already being provided the disabled and infirm elderly. The interviews were carried out in two districts of urban Guangzhou. She utilized a random sampling design to select the neighborhoods and informants to be interviewed. One innovative and promising technique which Ikels employed was the presentation to the elder informant and at least one other adult family member of brief scenarios involving cultural and normative dilemmas. She then asked them, separately if possible, to comment on the dilemmas (personal communication). This research should produce not only valuable data on the contributions of healthy elderly to the household and the patterns of care for the disabled but also thick descriptive accounts of dilemmas posed in the care of disabled elderly. Comparing the intergenerational differences within the same family will be especially provocative. (See Ikels 1988 for preliminary results.)

Yang Haiou is conducting sociological and demographic research on the effect of modernization on the well-being of rural elderly for her doctoral dissertation (personal communication). This study has four main objectives: (1) the predictions of the effect of demographic change in the last few decades on family size: (2) the exploration of the link between the well being of the elderly and family support and composition: (3) the examination of the effect of the new responsibility system on the family support network; and (4) the establishment of an empirical data base to be used in policy making for the elderly. She will use a microsimulation computer program to conduct family projections. During field research in rural villages in south China, she collected data for the quantitative and qualitative case study analysis.

A study on family change and the life course in Shanghai is being carried out by a group from the University of North Carolina, involving Glen Elder, Richard Udry, Gail Henderson and Anthony Obershall in collaboration with the Institute of Sociology at Shanghai University. This will be a survey research design involving random sampling to select a sample of 1200 respondents between the ages of 26-64. The study will utilize an event history methodology to be supplemented by a dozen "pop out" elaborations of important turning points in the life course such as first job, first marriage, migration, family formation, health crises, parents, sibs and children not in the respondent's household, sources of care for dependent household members, and societal events impact. It will also examine current household and family organization in such areas as family and household income transfers, division of labor, respondent's activities and chores, possessions; facilities and money management (Obershall, personal communication). By studying the elderly within the context of the family and the whole life course this approach is consonant with Chinese cultural notions of interdependence. Results from this research should enrich the study of cross-cultural comparisons of life course development.

     Studies directed at the family as a whole should help us understand better the context in which the elderly live and the forces which impinge on their well being and on the ability of their families to care for them. Janet Salaff and Burton Pasternak in collaboration with Pan Naigu and Ma Rong, from the Institute of Sociological Research, Beijing University, are conducting a study of families in three Han communities in Inner Mongolia. They will investigate the relationship between techno-economic setting and a number of sociological and demographic phenomena. During the first year a survey will be conducted of the three sites, to be followed by in-depth study of a sample of households and individuals in the next year. The project will address six general areas: (1) changes in technology and economy, (2) changes in domestic economy and their underlying causes, (3) changing patterns of marriage and postmarital residence, (4) changes in family form and composition as well as in the nature of intrafamilial relations, (5) demographic responses to policy innovations and to ecological and sociological diversity, and (6) patterns of, and factors affecting, migration (Salaff, personal communication).

## ANALYSIS

Gerontological studies in China are stimulating interest because of the opportunity they afford for significant cross-cultural comparisons, e.g. Goldstein and Ikels; Goldstein and Beall; Yang. So far, the most effort has been in the area of "modernization." There are also important opportunities for exploring the relationship of the elderly and their status within society to Chinese society as a whole and to specific aspects of that society, e.g. Davis's work on age stratification (Davis-Friedmann 1985a; Davis 1988). Ikels' work on the provision of and norms concerning long term care within the family is an example of this. There are many other opportunities for gerontological research in China which address issues specific to that context and in addition may have cross-cultural implications. The following represent some examples of research questions which might be developed.

In a singular fashion for a lesser developed country, issues involving the elderly have come to the forefront of national policy development in response to the one-child family campaign. Studying the process of developing policies designed to reassure people that their old age will be secure even without many children to care for them, provides researchers with the opportunity to analyze the effect of state policy on family life in a profound fashion. Unlike many government policy initiatives which affect one phase of the life cycle but inadvertently touch other phases, the current policy initiatives explicitly acknowledge the link between old age care and birth control and thus are designed to have a more pervasive impact on the whole life cycle of individuals and families.

The question within this context is, of course, will such an ambitious policy actually have an impact on the family? The debate over the specific impact of state policy on family life is already underway. In their work *Urban Life in Contemporary China* Martin King Whyte and William L. Parish (1984) argue that the rise in age at marriage and sharp declines in fertility are not distinctively an outcome of state policies but rather represent the process of change in family patterns brought on by industrialization, urbanization, and the other forms of social change which are associated with modernity. The anthropologist, Arthur P. Wolf (1986), disagrees with the Whyte and Parish interpretation and argues that, although natural declines in fertility are inevitable and explain some of the decrease in fertility rates, the rapid pace of the decline could not have occurred without significant government intervention. In her thesis on family life in Shanghai (1987), Dong takes a different tack and contends that the relationship between government and family life is an interactive one and that, while the family responds to state policies in areas such as birth control, it can also act independently by voicing concerns through the work unit and make demands on the state forcing new policy initiatives in areas such as child care and housing.

The formation of policy concerning long term care for the elderly is just beginning. In contrast to earlier times, the state appears to be interested in promoting family harmony. According to some policy makers the idea is to "...redefine the family as 'the basic cell of society', the image of more complex social relations, capable of integrating, or of serving as a bridge between the individual and society" (China News Analysis 1984). Studies which closely follow this development and people's perception of the policies and their response to them should contribute to an understanding of the intersection of individual family autonomy and state policies and to the place of the elderly in this system. Here Davis's (forthcoming) work on household interiors may provide an example of exactly how ethnographic research can contribute to the attempt to achieve a more comprehensive understanding of the place and role of the family in society. Davis reports that, with one exception, there was no evidence of political or work related themes in her informants' homes. This points to the clear primacy of the familial in people's lives, and of the inability of the state to control the content of popular culture. Ethnographic studies of this type are needed to understand the effect of state policy on the elderly within the family.

In line with an examination of the relationship between state policy and the family in the area of long term care, documentation of the singular cultural role played by the elderly in political life is significant. Davis-Friedmann (1983) describes how the elderly were utilized during political campaigns as witnesses attesting to the horrible life before the revolution and the substantial improvements following it. Under the Cultural Revolution, however, the elderly sometimes suffered as representatives of the old order (Olsen 1988). Currently the elderly are promoted as models of traditional values and as teachers of moral obligations and discipline to their married children of child-bearing age (China News Analysis 1984). It would be intriguing to explore how this unique political role for the elderly has affected their perception of personal and social generativity.

The impact of economic and social change on the extended family will be an important area of concern in developing an understanding of the status and well being of the elderly. This is closely related to the above topics but distinct in that it focuses on intergenerational relations directly and other larger forces only indirectly. Thornton and colleagues (1984, 1987) and Freedman's (1982) work on the extended family in Taiwan indicate that the family has proved flexible in adapting to the new economic and social conditions and in so doing has remained a source of strength and competitive advantage in the process of economic development. Based on research conducted in China, Croll (1987) describes an emerging family organization which she calls the "aggregate family" referring to separate nuclear units which continue to cooperate economically. Croll argues that the elaboration and strengthening of family networks is an advantage to the peasant family under the new responsibility system. This "aggregate" organization provides its members with access to capital for investments, labor exchange, employment, and urban ties. At the same time, she notes, the number of nuclear households has risen while the number of multi-generational, stem, and joint households has declined. If this is in fact a trend, it is not clear what the implications for the elderly will be. Perhaps they will live in separate residences but participate in an economically more efficient and interdependent family. Possibly in the short run, prior to adequate capital production, this could be detrimental to the elders. The situation could change for the better once sufficient surpluses are being generated.

The intense family focus does not bode well for those single and childless elderly who were dependent on the communal social welfare system. One suggestive development, which is not clearly related to the well being of the elderly but may affect them, is the desire of the successful peasant families to not call attention to themselves. Croll reports that rich households, to forestall criticism from neighbors and potentially the state if economic policies should change, are making an effort to integrate less fortunate kinsmen into successful enterprises. Possibly some social welfare benefit will arise from the same motivation. Village level as well as urban ethnographies will be important here in understanding family change and the impact this will have on the elderly.

Finally there is a clear need for more studies on such classic topics of ethnographic interest as mortuary ritual and ancestor worship. We know from the work of Davis-Friedmann (1983), Huang (forthcoming) and Parish and Whyte (1977) and Whyte (1988) that traditional funerals and ancestor worship are still practiced in some form, especially in the countryside, despite the prohibitions of the collectivization period. In what form these practices will continue and what effect smaller family size will have on them will be important questions for anthropologists as well as for the elderly themselves.

## CONCLUSION

Ethnography has an important contribution to make to the rapidly expanding field of Chinese gerontology. The large scale surveys which are currently being conducted or are in the planning stages will need ethnographic input both to ensure that the interview questions are culturally meaningful and to identify the specific cultural and social context of the respondents so as to determine their relevance for the population at large. Ethnographic studies will be important in their own right. Extended participant-observation studies are needed to document the process and impact of social and economic change. Longitudinal studies in the same village or urban district would be especially beneficial. Studies on purely ethnographic topics which document cultural practices, meanings and beliefs are needed.

State policy in China has had a profound impact on the position and role of the elderly and their experience of their lives. This has been accomplished through demographic policy, employment policy, and policies aimed at changing social and cultural life, in particular mortuary practices and ancestor worship but also patrilineality. The effect of these policies is still being determined and in fact remains in flux as the new responsibility system stimulates still more change in the social and economic structure of society. Although these changes affect everyone's life, analyzing them from the perspective of the elderly is particularly insightful because they appear to challenge a central Chinese cultural value – that of filial piety. Studying the elderly will thus provide a window into the larger question of what constitutes Chinese cultural identity in modern China.

REFERENCES

## I. English Sources

Baker, H. (1968). A Chinese Lineage Village: Sheung Shui. London.
Bannister, J. (1987) Implications of the Aging of China's Population. Presented at the International Symposium on Family Structure and Population Aging, Beijing, China, October, (1987).
Barclay, G. W. (1954) Colonial Development and Population in Taiwan. Princeton: Princeton University Press.
Cain, M. (1985) Fertility as an Adjustment to Risk. In Gender and the Life Course. A. Rossi, ed. New York: Aldine.
China News Analysis (1984) Socialist China, Social Policy and the Elderly. Number 1257, March 26.
China News Analysis. (1987) Working Women. Number 1334, May 1.
China Population Research (1987), 2 (1).
Cohen, M. (1976) House United, House Divided. New York: Columbia University Press.
Croll, E. (1987). The Aggregate Family: Household and Kin Support in Rural China. Paper delivered at the International Symposium on Family Structure and Population Aging, October 21-25, Beijing, China.
Davis, D. (1986) Family Supports for Chinese Elderly: Current and Future Trends. Paper delivered at the Annual Meeting of the Gerontological Society of America, Chicago, Ill.
Davis, D. (1988) Unequal Choices; Unequal Lives: Pension Reform and Urban Inequality. China Quarterly, 114: 223-242.
Davis, D. (forthcoming) My Mother's House. In New Ways of Studying Popular Culture in China. R. Madsen and L. Pikowitcz, eds. Boulder: Westview Press.
Davis-Friedmann, D. (1983) Long Lives: Chinese Elderly and the Communist Revolution. Cambridge: Harvard University Press.
Davis-Friedmann, D. (1985a) Intergenerational Inequalities and the Chinese Revolution. Modern China 11: 177-201.
Davis-Friedmann, D. (1985b) Old Age Security and the One-child Campaign. In China's One-Child Family Policy. E. Croll, D. Davin, and P. Kane, eds. London: MacMillan.
Diamond, N. (1969) K'un Shen: A Taiwanese Village. New York: Holt Reinhart.
Dong, B. (1987) Contemporary Chinese Intellectual Families. Doctoral dissertation, University of California-Santa Barbara.
Fei, H. (1949) Earthbound China. London: Routledge and Kegan Paul.
Freedman, M. (1958) Lineage Organization in Southeastern China. London School of Economics Monographs on Social Anthropology, 18. London: Athlone.
Freedman, M. (1966) Chinese Lineage and Society: Fukien and Kwangtung. London School of Economics Monographs on Social Anthropology, 33. London: Athlone.
Freedman, R. Chang, M. C. and T. H. Sun (1982) Household Composition, Extended Kinship and Reproduction in Taiwan: 1973-1980. Population Studies 36:395-411.
Fried, M. (1953) Fabric of Chinese Society. New Haven: Yale University Press.
Gallin, B. (1966) Hsin Hsing, Taiwan: A Chinese Village in Change. Berkeley and Los Angeles: University of California Press.
Gamble, S. (1954) Ting Hsien: A North China Rural Community. New York: Institute of Pacific Relations.
Greenhalgh, S. (1986) Shifts in China's Population Policy, 1984-86: Views from the Central, Provincial, and Local Levels. Population and Development Review 12: 491-512.
Greenhalgh, S. (1988) Population Research in China: An Introduction and Guide to Institutes. Center for Policy Studies Working Papers, Number 137. New York: The Population Council.
Greenhalgh, S. and J. Bongaarts (1987) Fertility Policy in China: Future Options. Science 235: 1167-1172.

Gui, S. X., L. K. Ki, Z. N. Shen, J. X. Di, Q. Z. Gu, Y. M. Chen and F. Qian. (1987) Status and Needs of the Elderly in Urban Shanghai: Analysis of Some Preliminary Statistics. Jounal of Cross-Cultural Gerontology 2: 171-186.
Ho, P. (1962) The Ladder of Success in Imperial China. New York: Science Editions.
Hsieh, J.C. (1985) Meal Rotation. In the Chinese Family and its Ritual Behavior. J. Hsieh and Y. Chuang, eds. Taipei: Academia Sinica.
Hsu, F. (1948) Under the Ancestors' Shadow: Chinese Culture and Personality. Stanford: Stanford University Press.
Hu, R., L. Wang, and Y. Zhang (1987) The Roles of Chinese Urban Elderly in Social Development: Direct and Indirect Participation. In Aging in China Family, Economics and Government Policies in Transition. Washington, DC: Gerontological Society of America.
Huang, S. forthcoming The Spiral Road: Changes and Development of a Chinese Village through the Eyes of a Communist Party Cadre.
Ikels, C. (1983) Aging and Adaptation: Chinese in Hong Kong and the United States. Hamdem, Connecticut: Archon.
Ikels. C. (1988) New Options for Chinese Elders. Paper delivered at a conference on Social Consequences of the Chinese Economic Reforms, Harvard University, May 13-15, (1988).
Ikels, C. (forthcoming) Family Caregivers and the Elderly in China.
Keyfitz, N. (1984) The Population of China. Scientific American 250: 38-47.
Kleinman, A. (1980) Patients and Healers in the Context of Culture. Berkeley: University of California Press.
Lang, O. (1946) Chinese Family and Soceity. New Haven: Archon.
Liang, J., E. Tu, and X. Chen. (1985) Population Aging in People's Republic of China. Paper delivered at the Population Association of America, Boston, MA, March 28-30, (1985).
Liang, J. and S. Gu n.p. Long term care in China.
Lin, Y. (1948) The Golden Wing: a Sociological Study of Chinese Familism. New York.
Liu, W.T. and E. Yu (1988) Informal social support systems for the elderly in Shanghai, China. Delivered at the Annual Gerontological Society of America Meetings San Francisco, November, (1988).
Nee, V. and H. Y. Wong (1985) Asian-American Socioeconomic Achievement: The Strength of the Family Bond. Sociological Perspectives 28: 281-306.
Nee, V and H. Y. Wong (1986) The Peasant Household Economy and Decollectivization in China. Journal of Asian and African Studies 21: 185-203.
Olsen, P. (1987) A Model of Eldercare in the People's Republic of China. International Journal of Aging and Human Development 24: 279-300.
Olson, P. (1988) Modernization in the People's Republic of China: The Politicization of the Elderly. The Sociological Quarterly 29: 241-262.
Parish, W. L. and M. K. Whyte (1977) Village and Family in Contemporary China. Chicago: University of Chicago Press.
Pasternak, B. (1968) Atrophy of Patrilineal Bonds in a Chinese Village in Historical Perspective. Ethnohistory 15:3.
Pasternak, B. (1986) Marriage and Fertility in Tianjin, China: Fifty Years of Transition. Papers of the East-West Population Institute, Number 99, July. Honolulu: The East-West Center.
Sankar, A. (1978) The Evolution of the Sisterhood in Traditional Chinese Society: From Village Girls' Houses to Chai T'angs in Hong Kong. Unpublished Ph.D. Dissertation, Department of Anthropology, University of Michigan, Ann Arbor.
Sankar, A. (1981) The Conquest of Solitude: Singlehood and Old Age in Traditional Chinese Society. In Dimensions: Aging, Culture, and Health. C. Fry, ed. South Hadley, MA: Bergin and Garvey.
Schultz, J. and D. Davis-Friedmann (1987) Aging China: Family, Economics, and Government Policies in Transition. Proceedings of the International Forum on Aging, Beijing, China, May 20-23, (1986). Washington, DC: Gerontological Society of America.

Tawny, R. (1932) Land and Labour in China. London: Beacon.
Thornton, A., M. C. Chang, and T. H. Sun. (1984) Social and Economic Change, Intergenerational Relationships, and Family Formation in Taiwan. Demography 21: 475-499.
Thornton, A., H. S. Lin, and M. L. Lee. (1987) Social Change, the Family and Well-being. Paper delivered at the Conference on Economic Development and Social Welfare. Taiwan: Academia Sinica.
Treas, J. and B. Logue. (1986) Economic Development and the Older Population. Population and Development Review 12: 645-673.
Tu, E., J. Liang, and S. Li. (1987) Mortality Decline and Chinese Family Structure: Implications for Old Age Support. Revised version of paper delivered at the Annual Meeting of the Population Association of America, April 30-May 2, Chicago, Illinois.
Whyte, M. K. (1988) Death in the People's Republic of China. In Death Ritual in Late Imperial and Modern China. J. L. Watson and E.S. Rawski, eds. Berkeley: University of California Press.
Whyte, M. K. and W. L. Parish (1984) Urban Life in Contemporary China. Chicago: University of Chicago Press.
Wolf, A. (1986) The Preeminent Role of Government Intervention in China's Family Revolution. Population and Development Review 12: 101-117.
Wolf, A. (1968) Adopt a Daughter-in-law, Marry a Sister: A Chinese Solution to the Problem of the Incest Taboo. American Anthropologist 70:
Wolf, M. (1972) Women and the Family in Rural Taiwan Stanford: Stanford University Press.
Yang, C. K. (1959) The Chinese Family in the Communist Revolution. Cambridge, MA: Harvard Press
Yang, M. (1945) A Chinese Village: Taitou, Shangtung Province London: Kegan Paul.
Yu, E., W. T. Liu, P. Levy, M. Y. Zhang, R. Katzman, C. T. Lung, S. C. Wong, Z. Y. Wang and G. Y. Qu forthcoming. Cognitive Impairment Among the Elderly in Shanghai, China. Journal of Gerontology: Social Sciences.
Yuan, F. (1987) The Status and Role of the Chinese Elderly in Families and Society. In Aging China: Family, Economics, and Government Policies in Transition. J. Schultz and D. Davis, eds. Proceedings of the International Forum on Aging, Beijing, China, May 20-23, (1986). Washington DC: Gerontological Society of America.

## II. Chinese Sources

Xia Lianbo (1978) Laonian Baojian Zhishi (Information for Perserving the Health of the Old). Shanghai: Kexue Jishu Chubanshe.
Yuan Jihui (1986) Laonian Wenti (Problems of Old Age). Shanghai: Fudan Daxue Chubanshe.
Yuan Jihui (1985, 1986) Chengshi Laonian Shenghuo Yanjiou (Research on Urban Elderly). Shanghai: Shanghai Laonianren Wenti Group. (Vol. I 1985, Vol. II 1986).

## LIST OF CONTRIBUTORS

Douglas E. Crews, PhD
Department of Preventive Medicine and Epidemiology
Loyola University Medical Center
2160 South First Avenue
Maywood, IL 60153

Bethel Ann Powers, PhD
University of Rochester
School of Nursing
601 Elmwood Avenue
Rochester, NY 14642

J. Neil Henderson, PhD
Suncoast Gerontology Center
University of South Florida Health Sciences Center
12901 Bruce B. Downs Boulevard
MDC-Box 50
Tampa, FL 33612-4799

Robert L. Rubinstein, PhD
Behavioral Research
Philadelphia Geriatric Center
5301 Old York Road
Philadelphia, PA 19141

Christine Fry, PhD
Department of Sociology and Anthropology
Loyola University of Chicago
Chicago, IL 60626

Christie Kiefer, PhD
Human Development and Aging Program
University of California
San Francisco, CA 94143

Andrea Sankar, PhD
Department of Anthropology
Wayne State University
Detroit, MI 48202

# COMBINED INDEX

Adjustment, 88-89
Africa, 21, 121-122, 136-138, 140, 143
Age grading, 121-122, 136, 142-143, 156
Age stratification 177
Alzheimer's disease, 24, 60-62

Biological anthropology, 11-38
Biology, 6-7, 11-38, 109-128

Chronology, 113, 129
China, 7, 46-47, 91, 156, 173-199
Clark, M., 45-48, 65, 153, 170
Confidantes, 115-116
Confucianism, 156-158, 164
Confusion, 77-79
Control, 87-88
Cross-cultural comparisons, 3-4, 21-22, 57, 69-70, 74, 78, 89-92, 129-150, 153
Culture, 1-7, 42, 44-45, 57-58, 60-61, 69-71, 74, 89-92, 109-128, 135-143, 155-158

Death, 4, 189-190
Dementia (see Alzheimer's disease)
Demography, 25, 176-177
Dependency, 165-168
Dougherty, M., 69-71, 74-75, 97

Environment and aging, 2, 83-84, 90

Falls, 81-82
Fry, C., 72, 97, 135, 139, 142, 147

Gender, 109-128
Gerontocracy, 119, 153
Gubrium, J., 55-56, 66
Gutmann, D., 117-121

Hazan, H., 53-54, 66
Health, 39-68, 77, 165-168, 185-186

Ikels, C., 135, 148, 174-175, 198

Japan, 7, 25, 140, 153-171
Johnson, C., 59, 66

Kayser-Jones, J., 57-58, 73, 75, 90, 99-100
Keith, J., 4, 72-73, 100, 135, 139-140, 142, 148

Life-course, 12, 15-16, 129-150
Longevity, 11-17
Long-term care, 4, 54-59, 188-189

Medical anthropology, 41-47
Medication, 79-80
Menopause, 12-13, 27
Mitteness, L., 39, 59-60, 67, 75, 101

Nursing, 69-109
Nursing homes (see long-term care)
Nutrition, 17-18
Nydegger, C., 120, 127, 132, 141, 149

One-child campaign, 177-178

Patrilineality, 121-122, 180-181
Power, 121-122, 124, 160, 164-165
Pressure sores, 82-83

Reciprocity, 182-185
Roles, 110, 141-142, 163
Rubinstein, R., 49-50, 67, 109, 124, 128

Sankar, A., 46, 68, 174, 198
Security, 160-161
Simmons, L., 4, 39, 130, 149
Social change, 158-160, 175-185
Social integration 161-164
Sokolovsky, J., 48-50
Stroke, 62-63

Time, 11, 87, 137-138, 141-142

Urinary incontinence, 80-81

Widowhood, 50, 116

Von Mering, O., 43-44, 68